Praise for *Seed to Seed*

ıtural History in its purest form. A botanical masterpiece in miniature'
ıvid Bellamy

Now that almost all of us spend our lives in cities *Seed to Seed* is just the ook we need to keep in touch with the living Earth on which, though we arely realise it, we depend entirely for our lives' James Lovelock

Tracing the life span of a weed, both in the wild and in a laboratory, nakes enthralling reading. Harberd, intensely aware of nature's changing easonal patterns, blends day-by-day observance with scientific exactness o sharpen our awareness of nature's purposes and beauty. Brilliantly written' Penelope Hobhouse

'Harberd leaves the sterile environment of the lab to get out in the natural world and bring back specimens that reveal the mysterious mechanisms of ife' *The Times*

'Harberd manages the difficult feat of showing how science can genuinely ncrease our sense of wonder about the natural world' *Daily Telegraph*

'*Seed to Seed* is a wonderful, original look at a field of research – plant genetics – that has received little attention from popular science writers . . . Any naturalist or gardener who has wondered what really makes plants grow will love this book. So will anyone curious about the process of scientific discovery' *Financial Times*

'Artfully composed . . . Harberd is brilliant at explaining the structure and growth of plants . . . *Seed to Seed* is a bravura performance . . . It is a privilege to watch a subtle and daring mind at work . . . and to glimpse, from so many angles, a scientist at work in the world'
Jenny Uglow, *Sunday Times*

'A unique and unusual book . . . both an enjoyable study of the life of a scientist and the kind of creative attempt to reach a broader readership that one hopes will be emulated . . . engaging and memorable'
Alan Packer, *Nature*

‘Mixing elegant explanations of gene function with impassioned observations of everyday plants, Harberd’s year-long diary explores their subcellular secrets . . . Possessed of a perceptive eye and a fine turn of phrase, Harberd has written a classic that will have you looking at thale-cress with all the amazement it deserves’ *New Scientist*

‘It’s exciting to look over a scientist’s shoulder while he’s at work, and Harberd’s enthusiasm certainly stirs the imagination’ *Scotsman*

‘Fascinating and original’ *Independent*

‘It’s another world. Another language. But he has the gift of connecting us with it . . . What Harberd’s book gives us clueless amateurs is a huge sense of awe at the extraordinary and brilliant machinery that regulates plant growth’ Anna Pavord, *Independent*

‘Lyrical . . . The joy and curiosity that carry sections of the book kept me enthralled from January through December’ *Seattle Times*

‘Delightful . . . Harberd gives us a very good account of plant structure and development in its more observable aspects’
Times Literary Supplement

‘It will tell you more about plant physiology and – in extremely practical terms – about the way science really works than anything I’ve come upon before . . . Harberd is sympathetic, a skilful and careful instructor, and almost without noticing you gradually realise that it is all making sense’
Literary Review

Seed to Seed

The Secret Life of Plants

Nicholas Harberd

BLOOMSBURY

First published in Great Britain 2006
This paperback edition published 2007

Bloomsbury Publishing Plc
36 Soho Square
London W1D 3QY

A CIP catalogue record for this book is available from the British Library

ISBN 9780747585619

10 9 8 7 6 5 4 3 2 1

Typeset by Hewer Text UK Ltd, Edinburgh
Printed in Great Britain by Clays Ltd, St Ives plc

All papers used by Bloomsbury Publishing are natural, recyclable products made from wood grown in well-managed forests. The manufacturing processes conform to the environmental regulations of the country of origin.

www.bloomsbury.com

For Jess

. . . what greater delight is there than to behold the earth apparelled with plants, as with a robe of embroidered worke, set with Orient pearls and garnished with great diversitie of rare and costly jewels? If this varietie and perfection of colours may affect the eie, it is such in herbs and floures, that no *Apelles*, no *Zeuxis* ever could by any art expresse the like: if odours or if taste may worke satisfaction, they are both so soveraigne in plants, and so comfortable that no confection of the Apothecaries can equall their excellent vertue. But these delights are in the outward senses: the principal delight is in the mind, singularly enriched with the knowledge of these visible things, setting forth to us the invisible wisdome and admirable workmanship of Almighty God . . .

John Gerard, *The Herball*
or General Historie of Plants, 1597

Contents

Preface 1

January 3
February 15
March 43
April 97
May 127
June 159
July 183
August 209
September 227
October 259
November 281
December 295

Afterword 301
Glossary 303
Acknowledgements 310

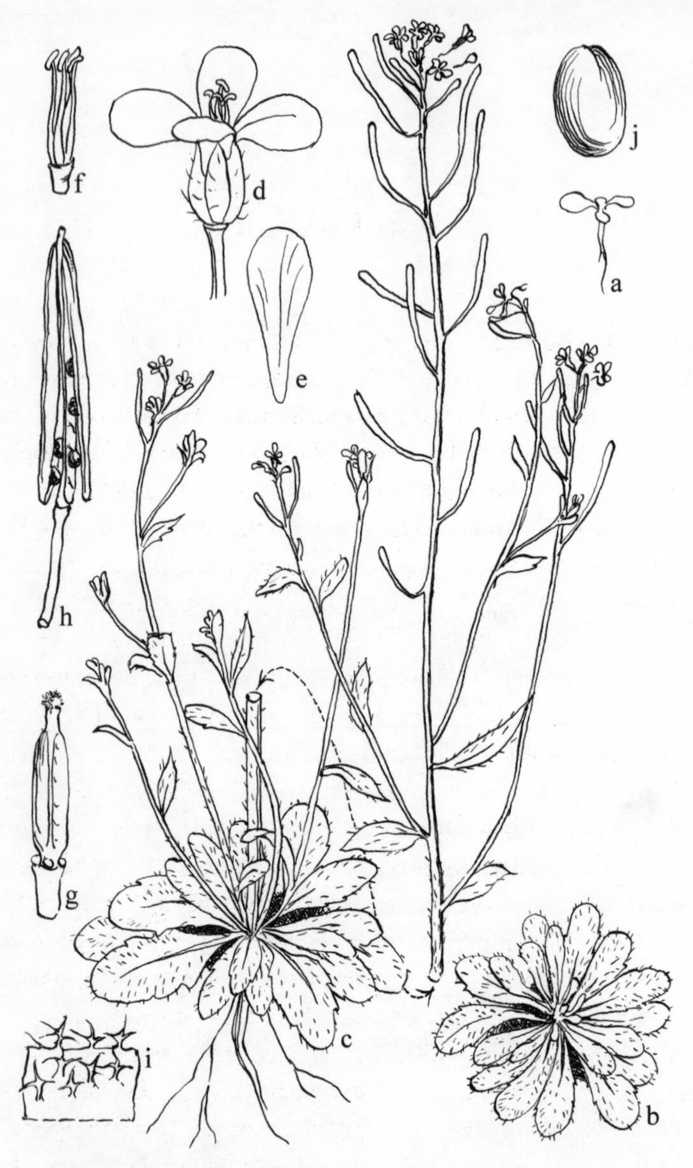

Arabidopsis thaliana (Thale-cress).

a, a seedling; b, plant at vegetative rosette stage; c, plant at flowering stage; d, flower; e, petal; f, flower with sepals and petals removed; g, gynoecium, nectariferous glands and upper part of pedicel; h, opening seedpod; i, upper surface of leaf showing trichomes; j, seed

PREFACE

THIS is a book that attempts to show how science can enhance our vision of the world. It is a book written principally for non-scientists. On one level it describes the wonders of the area of science with which I am particularly engaged: the developing understanding of how and why plants grow. On another level it depicts these recent advances within the context of a wider representation of mind. That mind is my own, occupied as it is with some of what I think to be the most exciting current questions in plant biology. The book is, therefore, in part, a mental self-portrait. Like all portraits, it abstracts: it exaggerates some features, omits others. It is drawn in the hope of revealing things that were previously not visible.

I am a scientist who is fortunate to be working at one of the world's foremost scientific research institutions. I direct a research team at the John Innes Centre in Colney, on the outskirts of Norwich. Our experiments are revealing the hidden fundamentals of how the growth of plants is controlled. Finding the genes and proteins that build plants into the visible shapes and forms they have. Exploring the intrinsic beauties of these invisible things.

The book is really a notebook, a diary of the year 2004. At its core is the progression of a chosen plant through the stages of its life-cycle, set within the context of the seasons. This description is combined with an outline of the unseen molecular forces that drive plants from stage to stage. Other progressions are also recorded: the deepening of scientific understanding of the growth of plants (a history of the recent work of my research group); the realisation of a new research direction; a personal journey from frustration to enlightenment.

What more needs to be said by way of introduction? Perhaps that much of what follows consists of notes and sketches rapidly executed in snatched moments. Often neither finished nor complete. Sometimes half-thoughts. My intention was to try to capture a sense of the scientific process within a more general picture of a mind that is engaged with it. Feelings are recorded – and feelings are things which we scientists are often overly constrained in expressing. Most of all, I hope that I have occasionally managed to evoke a sense of life. The passing of a moment in a few dashed jottings.

N.H.
January 2006

JANUARY

Thursday 8th January

EIGHTEEN years ago today, I first flew out from London to California. A new beginning. And, for someone who had lived most of his life in England, a huge change. So many things were different. The quality of the light, for instance. It penetrated, had a revealing clarity I'd not previously seen. I have splintered memories: a glimpse of the Golden Gate bridge, a great ochre harp shimmering in the wind-flecked sea; sudden alarm at a brief, ground-shuddering earth-tremor; violent rain in a storm from the Pacific.

Whilst I was in California I absorbed new styles of thinking, new ways of doing science. Things that seemed to have grown from the place, from the landscape and from its people.

I am a plant geneticist, and had gone to California to begin work in a new area. As well as swapping continents, I was swapping plants: from wheat to maize. And I was quickly captivated by the bold magnificence of the maize plant, the vigour of its growth. With a summer crop in California and a winter crop in Hawaii we had two generations a year: six months in between to plan new crosses and consider the outcomes of the previous ones.

It was an exhilarating time in plant genetics. I worked long hours with enthusiasm, returning from the heat of the field drenched in pollen and sweat. Around the world, other groups were using a variety of different plants for genetic research: wheat, barley, rice, tobacco, even snapdragons. New concepts were coming from all directions, new areas of research were continually opening up in front of us.

And then, on top of all this, there began at that time a shift in thinking that was eventually to propel plant genetics on to an entirely new plane. This shift was based on a unifying idea. An idea rooted in the concept that all plants are essentially as one, that the different species of plant have more in common than they do that divides them. That cacti, tree-ferns, redwoods, oats and sunflowers are more similar than different. The idea was that concentration of effort on the study of a single species would advance understanding of all.

As acceptance of this idea grew, so the next question arose: which single species? In the end, the species that many plant biologists settled on was the thale-cress: *Arabidopsis thaliana*. Thale-cress? What is thale-cress? You've probably never heard of it. A squat rosette of leaves that bolts to a height of a foot or less, flowers, then dies shortly thereafter. Scattered about in the neglected parts of gardens, in wasteland or on walls. Inconspicuous and ignored; common, but unfamiliar.

So why was thale-cress preferred? Because it has attributes that are perfect for the plant geneticist, making it our own *Drosophila* (fruit-fly). First, it's easy to grow in glasshouse and laboratory, its small size allowing many thousands of plants to be grown in limited space. Second, it has a relatively short generation time: seed to seed in six weeks in the lab. That's eight generations a year, rather than the two with maize. Four times as many new crosses every year, each one bringing new insights, deeper understanding. Third, thale-cress has a relatively small genome, a property of great potential to a geneticist. Determination of the DNA sequence of the entire genome, the totality of the DNA that contains the genes, was therefore a more realistic proposition than for other species. It would be the first sequenced plant genome. An exciting prospect since the DNA is the key to understanding how a plant grows. If we knew the sequence of all of the genes of the thale-cress we could really get to grips with solving some of the most important questions in plant biology.

So I returned from the brightness of California to life in Norfolk, with my mind enriched, with a livelier, more creative approach to science. Swapped plant again, from majestic maize to humble thale-cress, and am glad to have done so. In the past decade, the unifying idea has been realised with spectacular success. The genetics of thale-cress has been developed to a level of huge sophistication. We now use this genetics to probe the hidden secrets of plant life. Thus revealing knowledge of things as diverse as the formation of flowers, the germination of seeds, even the mysterious but familiar process of growth itself (my own area). The recently completed thale-cress genome sequence has revolutionised our approach: telling us about new genes we didn't know were there, sharpening our approach to our experiments, giving sight of something fundamental to life itself.

So the idea worked, and continues to work, with a degree of success I doubt even the most enthusiastic of its original proponents could have envisaged. Through the lens of thale-cress, we can see deep into the lives of the plants we encounter in our everyday lives.

It's been exhilarating to have been involved in all this. The momentum of the progress is such that I know it will continue for some time to come. Yet within my own part of it I've come up against a barrier. Where next?

Friday 9th January

In the last year or so the work of my research group has been going well. In particular, in 2002, we experienced a renaissance. New ideas sprang to mind. Subtle experiments tested these ideas. As a result, an important discovery was made. This resurgence reached its peak in the spring of 2003, when we published a paper describing that discovery.

And although the excitement was sustained for several months thereafter, towards the end of 2003 I began to experience a sense of unease. Associated perhaps with the failing of the light at the approach of winter. I was becoming aware that I couldn't see where to go next. That I had no sense of how to advance further.

Of course science is always like this. There are peaks and troughs. I've experienced both. But the problem with being in a trough is that it is a place from which the view is limited. There is the feeling of being trapped with no way out. And always the question of how long the entrapment will last. A self-sustaining state: at the time when new vision is most needed, it is most unlikely to come.

Monday 12th January

I've been trying to think. Attempting to tighten my strings. On Saturday I walked around the cathedral. Making spokes in my mind to the top of the spire from the different points on my progress. Shaping some new ideas. But, as a friend once said to me in a bar in County Cork, 'You're a man of ideas, Nick, and the trouble with ideas is that they don't work.' On the way home, my confidence in the structure collapsed. I'm back where I began.

Tuesday 13th January

What has the weather been like today? I don't really know. I've been so busy with running the lab that I was scarcely aware of it. I think it rained. And certainly, when I went to collect Alice and Jack from school, the playground was wet. Dirty puddles on the asphalt.

Whilst walking home, I asked what they had been doing. There was the usual flood of eloquence from Alice. Then, from Jack, bored: 'Beans and stuff.' And in response to my further prods: 'You know, Dad, the usual thing. Same as Alice did last year. Beans in a jam-jar. We're going to watch them grow.'

Suddenly I remembered a moment from my own childhood. It was April, perhaps May. A sparse classroom at my infant school. High ceiling, bare floorboards. A beam of sunlight filtered through eddying specks of dust, falling on some twigs of horse chestnut standing in an old jam-jar of water. The day before, they'd carried large, fat buds which I'd squeezed, feeling the plump firmness and the sticking tug of the gum on my fingers. Now, as I ran into the room, I instantly saw that the buds had been replaced by fragile lime-green leaves, leaves that were expanding into the warmth and the sunlight. I stopped running, stood still, and took the whole thing in, letting my gaze travel up from the scarred, pitted grey wood, past the splayed and shrivelling scales of the bud-cases. To the apex of the thing, an ultimate chorus of green furry arms and the outstretched hands of leaves. The hands, tiny miniatures of their final form, laced with veins, reaching to the sun with expanding palms and fingers. Shining with the light of miracle.

This is the kind of vision I need. For the past few months I've been doing little other than going back and forth from home to work. Perhaps I should get out into the real world more than I do. See something else.

Friday 16th January

Last night the temperature plummeted to well below freezing. I awoke cold before dawn, although I didn't feel it as cold. More as a pressure on my chest, a restlessness in my arms and legs, faster breathing, heart thumping at my ribs. Passing my hand over my temple, I felt my clammy skin, fingertips wet with sweat from my scalp as they raked my hair. The thought that the earth was sucking the warmth of my life out of me.

This morning, the lawn was white, grass blades etched with the grain of frost. Sun faint through fog. I'd decided to take a day away from the lab. To get on my bike and go somewhere else. I wasn't sure where, just somewhere – out in the fens, towards the broads, or perhaps along the old railway track to Reepham.

But I didn't get very far. At the end of our street is a slight slope down to the junction with the Unthank Road. It's shaded, prone to icing in cold weather. I came down too fast, braked to avoid a car waiting at the junction, skidded, lost balance, fell hard on to the tarmac. Although I was unhurt, the fall unnerved me. A sudden blow that divided time into what went before and what followed.

Monday 19th January

Alice's mind has been sparking. Thinking about Jack's beans, knowing that I work with thale-cress, she asked if it too grows from beans. So I brought some thale-cress seeds home from work to show them to her and Jack. They were amazed at the seeds' tiny size. It *is* astonishing that an adult thale-cress plant, which can attain a foot or so in final height, originates from such a minute salt-grain-sized brown dot. Viewing them through the old low-powered microscope I keep at home, Alice and Jack saw that thale-cress seeds are indeed like tiny beans, rounded and plump. As I watched my children looking through the microscope, oscillating between wonder at what they saw and scraps over whose turn it was to look, I thought about the affinity between the inertness, the dry dormancy, of those seeds and the winter season.

Jack is proud of his bean. He says he can see the beginnings of a root growing down and a shoot growing up. His telling me this evoked another childhood memory. Of a long-ago cold January afternoon, in failing light. I stood, a small figure in the wind, my roundness exaggerated like a deep-sea diver, arms and legs fat with padded clothing, feet in boots, my face peering out of a hood.

I looked over the garden, beyond my father, and out to the trees at its edge. Above them the sky was in broad layers of cobalt cloud, edged with orange where the low sun glowed on a fast-approaching bank. I looked back at my father, at the orange light on his coat. He was working fast, loosening the soil around a row of parsnips. His coat was open, and I could smell the sweat and heat of his body. He stabbed at the ground with his fork, rocking it side to side. Then he used the fork as a lever, his right hand on the handle, his left making a fulcrum on the shaft, the fork end under the parsnips. He worked along the row, from plant to plant, then stuck the fork into the ground to one side. Bending over and grabbing at the browning foliage, he yanked a parsnip out of the earth, jerking at it as a blackbird pulls a worm from the ground, the worm first curved, then straight and taut. He lifted the parsnip into the air, holding it by the old leaves at its crown, looked momentarily at the dirty yellowish cone of the root, and then, looking towards me and smiling, dropped it on the ground close to the hole it had left.

I wanted to do it too. It seemed like unwrapping a present. I tottered towards him, unsteady over the rough furrows, and tugged at his trouser leg. He stopped what he was doing and watched whilst I pulled fruitlessly at another mop of brown leaves. He laughed. Not cruelly but with happiness. The high-pitched staccato laugh of an emotional mind. He took his fork again, loosened the soil some more, gave a preparatory tug to the leaves, and then stood back whilst I pulled again. Suddenly the parsnip came out of the soil and into my lap as I lost my balance and found myself sitting on the cold ground next to the empty hole.

I looked closely at the parsnip in the fast-fading light. This one was forked, the root tapering to two narrow cones, with smaller hairy roots growing out of it. Crumbs of soil stuck to the surface and the hairs. At the top of the parsnip, close to where the surface of the soil had been, the root had a flat top out of which grew the browning stems that carried the tattered remains of a few leaves. My mind drew a line that represented the soil's surface, separating leaves and shoot from root, dividing above from below ground. The line then expanded its dimensions to become a plane, an extended flat surface, that divided all the other parsnips still in the garden into root and shoot. The plane was parallel and coincident with the surface of the soil. Knowing, without knowing why, that the plane was important, I bent forward and began to peer through it into the hole. I looked at the rough, dark walls sloping into the earth, at the curly dead roots and old twigs that grew out of the sides, and followed them down into dark invisibility.

Then I looked up, into the light, back on my side of the plane, and saw my father looking thoughtfully into the sky at the approaching bank of cloud. As I saw him standing there, it seemed to me that the parsnips lived in two worlds, and came in two parts, divided by my plane, the shoot above the ground and the root below.

Then my father said something about leaving what was left of the parsnips in the ground. And suddenly, as the pace of the light's fading intensified, hard hail stung my face, white stones bouncing in the furrows of the garden like seed-corn thrown from a sower's hand. My father picked me up and ran for the shelter of the trees. But I was still thinking about my line.

Thursday 22nd January

WHEATFEN

It's milder now. A few days ago the ice retreated in the face of warm air brought west from the Atlantic. Over Ireland, the Irish Sea, Wales, the Midlands and now here. And although the warmth brought rain with it, the rain has stopped, leaving a clouded grey sky.

Today I discovered Wheatfen. Some months ago, a friend told me about the existence, somewhere near Surlingham, of an area of reed-bed and fenland that had been the home of Ted Ellis, the celebrated Norfolk naturalist. Now the land is a nature reserve, maintained by the Ted Ellis Trust. I'd never been there, and this morning I decided to go.

Exactly where Wheatfen was, I didn't know. And it wasn't easy to find. I cycled past Surlingham St Mary's, took the left turning up Pratt's Hill, thinking this would bring me close to the river, alongside which I knew Wheatfen to be. But it became clear that Pratt's Hill was a mistake. I retraced my path, turned through the main street of Surlingham, passed the duck-pond, took a left turning, and finally noticed a small sign directing me down an unmade road.

Wheatfen is simply glorious. I walked around it, taking it in. A huge expanse of water, reed-bed, fenland and wetland, stretching into the river Yare. Bounded by woodland. I cannot begin to encompass it in description. It is about 130 acres in extent, too intricate in detail to cover in a few words. Yet I can point to the individual features that struck me today. The sense of space – distant flat horizon and massive sky. The wetness. The colours – dun reeds, grey clouds. The occasional scurrying moorhen.

I'll go back to this place.

Saturday 24th January

We went to the Theatre Royal to see *The Play What I Wrote*. It was fun to see these representations of Morecambe and Wise, comedians who had brought such delight in my childhood. Moving to be part of an audience experiencing collective stirrings of memory. Remembered catch-phrases and actions that meant so much thirty years ago. All re-formed now in the telling, making a new thing from the old.

This is the way to arrive at a new direction. By seeing things already seen in a new way, making predictions of unseen things, testing them. But how to do it in reality?

Wednesday 28th January

I've had no time to write the last few days. Extremely busy: meetings, discussing latest results with the team. Also, we've been rushing to complete a manuscript to submit for publication.

Seems to have taken so long to move from initial disconnections and lumps to something that weaves text and figures into an integrated whole. Nearly there now.

Thursday 29th January

I wish I knew what to do next. I need to define a question. If I can work out what the next question is, the path will become clearer. But thinking of questions, the right questions, is *so* difficult. And will power is not, in itself, sufficient.

Friday 30th January

The cold has returned. Walked the children home from school in falling snow. They scampered about, chasing flakes, catching them, melting them in their mouths.

Jack is delighted with his beanstalk. Enjoying the attention. Because, strangely enough, his stalk is growing faster than all the others in the class. Was his a magic bean? What made it grow so tall? Who knows – the convergence of many things perhaps: its genes, its position in the light, its watering, his care for it . . . the multiplicity of factors that might collectively be contributing to its growth is infinite. He asked me why it grows so tall. Found it funny that I, an expert in plant growth, couldn't tell him.

FEBRUARY

Tuesday 3rd February

STRANGELY mild today. A rebound from last week's ice and snow. The wind in westerly gusts. Sky flat, uniform grey. Bubbling blackbird song soothing the frustrated mind. In the garden, the yellow flowers of aconites splayed open by the warmth. Hanging snowdrop pearls, split three ways. Crocus spikes erect.

Wednesday 4th February

The mildness continues, extreme for February.

I'm a biologist because I love life. Although I occasionally feel uneasy with the sense of division that the term *biology* evokes. If it didn't sound so pompous, I might prefer to think of myself as a natural philosopher. At least that is less constraining in its resonance.

But right now I lack the philosopher's vision.

Thursday 5th February

A day off. I got on my bicycle and returned to Wheatfen. The wind still south-westerly: the mildness continues. Although the cold is shortly to return, according to the forecasts.

In Norfolk the sky yawns. Today layered clouds dominated the landscape. At the highest level a buff carpet, threadbare with blue holes. Below, individual dumpling lumps with grey rounded sides and dark, flat bottoms. The sky a salad of dampness: greys, blues, and yellows, all speeding in one direction.

On the ground, light shining then fading, crossing from west to east in fast, bright patches. The wind striking in moist gusts. The swirling of damp brown leaves. Changing scents: sweet, wet beech woods, sheep-dung, my warm sweat, the sewage-stench at Whitlingham.

As I cycled, a thought pushed through to the surface of the mind. That there were just a few miles of turbulent air and vapour above my head, and then the vast nothingness of the universe. In the thought I could see a vector. It began at the centre of the world, passed through my feet and head, and then stretched into infinity. I felt small. Exposed out there on a bicycle in the wet wind. Part of a thin sandwich-layer of life, with the empty heavens above and the earth's molten core below.

In Whitlingham lane, my glance alighted on a leaf amongst the brambles lining the road. Its mottled green hand spangled with bright droplets from an earlier shower. There was a flash of recognition. That the life of the leaf, of its cells, vessels, and hairs, was linked with my own. Hardly a great revelation. Surely it's clear to everyone that life relates to life? But there was a peculiar intensity about that small epiphany. A momentary certainty that that blackberry leaf and I belonged to a single entity. And it was then that I thought of looking for thale-cress.

Of course I see thale-cress growing every day, in the greenhouses and the lab where I work. With the idea to look for it growing in the wild came the thought that perhaps it was strange that I had never felt the urge to do so before. That I've spent so many years of my life researching the growth of a plant I've never sought in nature. That here was a disconnection. The thought echoed that of a few days ago: that perhaps I've spent too long amongst computers, microscopes, and test tubes. Perhaps it's time to get out.

So I began to look, stopping occasionally along the way. At what seemed to be likely sites. Gritty verges. The margins of ploughed fields. The sandy edge of the track that runs over the rise from Whitlingham to Wood's End. Cracks in flint walls. But I found no thale-cress.

Then, when I reached Wheatfen, I walked for a while amongst the flooded reed-beds. Was briefly distracted from my search, looking at the angles with which the straight thatch stuck up from the grey ruffled water. At an inclined plane, flattened by the weight of last week's snow. Of course I knew this was not the place to find thale-cress. It doesn't usually grow in water-logged soil. So I went into the wood, where the soil is drier. And although I felt momentary exhilaration amongst the bare trunks and waving branches of the blustering afternoon, I didn't find thale-cress there either. Nor at the margins of the reserve, in the car-park, or on the track.

Later, at home, I leafed though *Flora of Norfolk*. It read: '*Arabidopsis thaliana*, Thale-cress: Summer and autumn germinating annual, abundant in open, dry soils, on waste ground and on walls. Those germinating following June rains may flower in September and October. On heavier soils and in Broadland, often confined to wall tops and gravel paths.' And there was a picture, a photograph of a thale-cress plant growing in the gravel on the surface of a grave, a tombstone behind it.

Friday 6th February

Cycled to John Innes in persistent rain. Drops exploding in ring-splashes on striking the road.

The paper we wrote and submitted a few weeks ago was rushed because we'd learned that a competing group had submitted a manuscript for publication describing experiments and conclusions we believed to be identical to some of our work.

I'm sure that both parties will produce good papers in the end. But as always when faced with the reality of scientific competition I'm feeling uneasy. Where does the responsibility for this awkwardness lie? Is it personal? Or is it a general responsibility of the scientific culture, of the way we scientists do things?

Saturday 7th February

The crocuses in the lawn have shot up like rockets in the recent mildness. But today the cold is returning. The sky blue with patches of high cloud and haze, the light crystal-hard, wind squalling, bending trees. A pair of magpies fly up and then stall, are stationary for a moment, riding a gust.

I'm developing an idea. The seed planted in my mind by Jack's bean. That I should continue the search for a thale-cress plant growing in the wild. Then, when it is found, I'll record the growth and life-history of that plant in these pages. The same plant that I've studied so closely and for so long in the lab. Perhaps this will rekindle my sense of wonder, will help me to see the way forward for our stalled research. It will be a new natural history.

I have of course been a lab scientist all my life, not a field naturalist. More accustomed to controlled experiments than to observing and recording the events of the natural world. Let's hope that I can learn as I go. Clear description of what is seen is the key. Illuminated by an account of the unseen events that drive those observed phenomena.

In essence, the idea is to represent nature's progress. To chart the life of a chosen plant, through spring, summer, and autumn. To observe that precarious passage, the stages of a life-cycle, an eventual death. To see a new beginning in the succeeding generation. This is an ancient story, often told. But I think I can tell it in a different way. I've been spending my life shaping a vision of hidden molecular events, invisible things that drive life. I think it's time to retell the old story in new terms.

A natural history. The idea reminds me again of school days – the nature table, horse-chestnut buds, and so on. So who do I imagine will read this natural history? Me, yes, but perhaps – at least this is my growing hope – Alice and Jack. Not now, they're too young to stay with it. But in a few years time, I'd like it if what I wrote here enabled them to see more clearly what was in my mind than might have been possible had I not written it.

Monday 9th February

Last night, a roaring gale driven by a strong north-westerly wind. Lightning. Thunder. Hail banging and clattering on the window. Lying in bed with pleasure in the simultaneity of snug warmth and wild noise from outside. The contrast heightening the enjoyment of shelter, of the protecting roof and walls. And I admit to a perhaps childish excitement in the spectacular aspects of extreme weather.

Tuesday 10th February

FINDING A PLANT

A cold, bright, blue morning. Very still. The sky criss-crossed with aeroplane vapour trails. At home, the wood pigeons cooing. I couldn't resist the temptation to cycle out on such a day. To Wheatfen first, then on in search of thale-cress. And I soon found myself striding through brown reeds and frosty bracken in the brilliant light. Wrapped in my tweed coat. Imagining myself for all the world like some Victorian naturalist-clergyman, glorying in the intricate beauties of Creation.

After Wheatfen, I backtracked, visiting St Mary's church and its surrounding graveyard. The search for thale-cress continued, prompted by the photograph I found last Thursday.

I rested for a while in the sunlit shelter of the flint wall that marks the boundary between graveyard and road. Then, walking along a line of graves to the north of the church, I suddenly felt the excitement of anticipation. Amongst the gravel covering one of these graves, I could see small splashes of green.

I walked closer, squatted at the foot of the grave. There is a headstone in the form of an open book. A body-length strip of ground. A low surrounding wall of mottled cream marble, the stone oddly cracked near one corner and patched with brown, like tea splashed on paper. The gravel is thinly laid, revealing in places the dark, wet soil beneath it. A heterogeneous landscape of aridity and moistness. Scattered amongst the sparse chips of stone are diverse weeds: docks, dandelions, thistles, and so on. And amongst these, some thale-cress: star-like rosettes of leaves studded in the gravel.

The search was over. What I'd found was a curved line of three thale-cress plants. All three of about the same age, probably seeded last year. Plants like those I see every day in the lab, but these were not cultivated, not deliberately planted.

Sketch-map of grave.

The first thing to do was to choose one. So I did. The middle one of the three. A battered, torn rosette: fragile outer leaves surrounding an inner verdant spiral. This plant would be *my* thale-cress. The object of my observations. Its progression through life recorded in my notes. As I chose it, I experienced a transient, tugging thrill of apprehension. Thale-cress is a plant that excells at the precarious. It seeds itself in marginal ground, on walls, on sites where the supply of water is chancy. Places that veer from aridity to flood. Thale-cress is equipped to meet such challenges. But it is always living on the cusp of risk.

To begin with, then, a description of this chosen thale-cress. To be written as though I've never seen one before. Where to begin? First I'll try a longer-distance view, then I'll describe the plant's appearance from close up.

Standing up, backing away from the grave, at a distance of a few feet, what struck me most about this plant? First, its colour. Its greenness gleams against the grey gravel and dark soil. Even from this distance the gleam is not uniform. The younger leaves and centre have a deepness, a blueness, to the green that the older, yellower leaves lack. The greenness of each leaf having a marbled texture that extends across the flat plane it presents to the world.

The next thing that struck me was structural. That the plant is built around axes of symmetry. Take the leaves, for instance. Running through the centre of each leaf, from tip to supporting leaf-stalk, is a pale line, almost yellow, that divides the leaf into symmetrical halves. And symmetry shapes the whole plant, defines the arrangement of the leaves. Seen from the side, the thale-cress is a tiny bush, standing just proud of the soil. The lower leaves form a shallow arch, with tip and stalk base both touching the soil. The upper leaves are held more into the air, making the bush symmetrical around a vertical line. Moving closer, looking down on the thale-cress, a further symmetry becomes clear. It is in the form of a star, a rosette. The leaves radiate, pointing tip-outwards from a central point. A circle could be drawn around the tips. I thought of Leonardo da Vinci's drawing of a man with feet and hands touching the circumference of a circle. As in this drawing, the thale-cress has radial symmetry, a centre and a circumference.

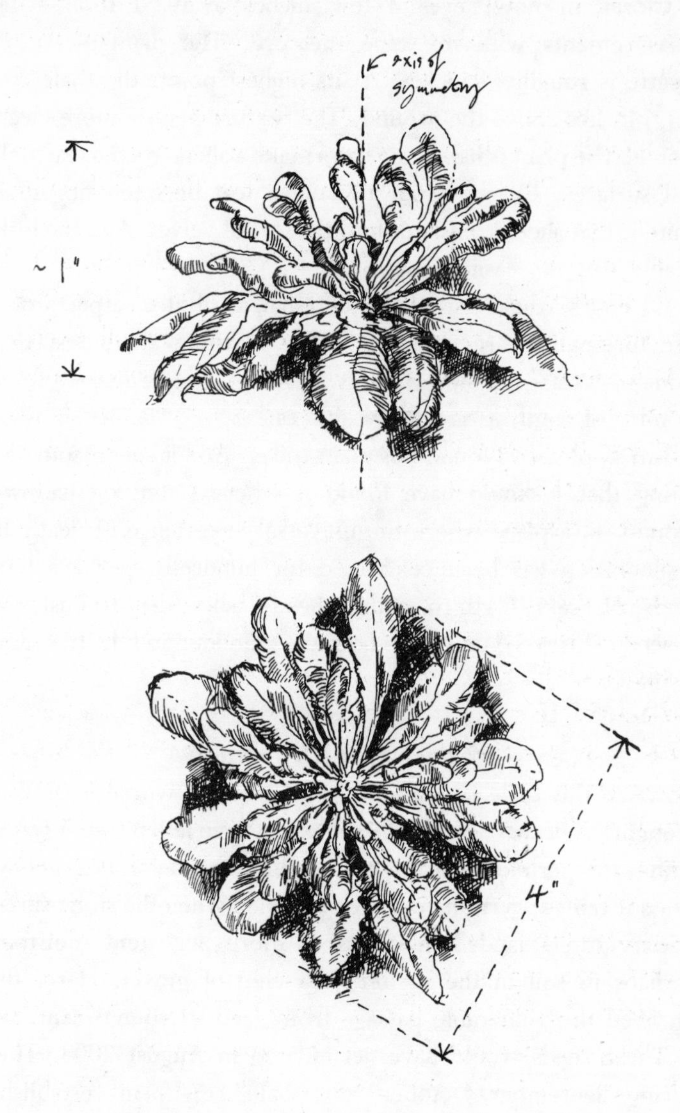

The thale-cress plant from the side and from above.

Closer in now, eyes a few inches away, I took a few measurements with my tape measure. The diameter of the rosette is roughly 4 inches. At its highest point, the thale-cress is 1½ inches above the ground. The texture is now more clearly visible. The plant is hairy: the hairs make a sheen on the leaf-stalks and surfaces. Reaching out to take a leaf between finger and thumb, that sheen translates into the feel of velvet. And the leaf is pliable, supple, easily bent.

Of course thale-cress has additional attributes, aspects of its structure which are invisible. These invisible attributes, these hidden secrets, are nevertheless as real, as contributory to the wonder of it all, as are the visible ones.

I'm so pleased I found this plant today. And it seems somehow fitting that I should have found it where I did, on hallowed ground, in a place where the universal experience of death and regeneration has been celebrated for hundreds, perhaps thousands, of years. Earth to earth, ashes to ashes, dust to dust, seed to seed. This weed embodies both the grandeur and the transience of life.

Wednesday 11th February

How did this thale-cress come to be growing where it is? What brought it here? A scatter of seeds must have fallen to the ground within the perimeter of the grave. Some fell on stones, got wet when it rained, germinated, then perished when the stone surface dried. Others landed in places of more persistent moistness, perhaps in soil in the shelter of a chip of gravel. Here, they survived the vulnerable passage from seed to young plant.

These seeds would have germinated in August 2003. Then, during September/October, the thale-cress plant established itself as a young seedling. With a small leaf-rosette expanding across bare earth and grit, and roots tunnelling into the earth below. In November, the growth of the plant first slowed, then halted as the cold intensified. Now, in February of this new year, its growth is beginning again.

But how that seed came to fall on the grave is a mystery. Especially since yesterday, before leaving the churchyard, I walked around it a little. Took in the scene, noted the fine trees that mark its edge. Looked at other graves. And nowhere did I find another thale-cress. The ones found are seemingly unique to the area. I'm lucky to have found them.

Thursday 12th February

ON GENES AND SENSITIVITY

It is still. Grey and cold. But there was a real sense of progression this morning. A sense of winter ending and spring beginning. Our northern hemisphere tilting closer to the sun.

After work, went back to look at the thale-cress. The winter's history is written in its leaves. The oldest, hidden beneath the bush of the rosette, are faded, brown and wilted. Lying on the ground, unable to support their own weight. The oldest leaf of all is dead, clings like a shroud to the soil's surface.

The next-oldest leaves have broad yellow edges, and a green centre. Whilst the centre is still alive, the edge is dying. The green is the colour of chlorophyll, the pigment that absorbs the sunlight that keeps the plant alive. The leaf edges are yellow because the chlorophyll has been salvaged from them for reuse elsewhere in the plant. These aging leaves are extensively damaged. There are holes, gaps, areas eaten by slugs, snails, or insects.

This thale-cress plant was exposed to the full severity of winter. Blasted by gales. Struck by hail. Seared by frosts. Yet it survived. And despite the fact that this happens all the time, it remains astonishing that so vulnerable a thing survived in the face of such forces.

The secret to the plant's survival is its exquisite sensitivity to the world around it. It senses, then responds to, and hence survives, adversity. This sensitivity is given to the thale-cress plant by its genes. So its genes got it through the winter. Battered, ragged-leafed, and scarred, yes; a close-run thing, quite possibly; yet get through it did.

I suddenly realised that dusk was falling, that I had to go home. I was ready for the exercise. My legs restless. Almost as if they were willing their own movement. I'm excited by this new project; the urge to move is a feeling that commonly arises when new ideas are welling up. The first time for months that I have felt like this.

And on the journey home I imagined a merging of the landscape of fen and wood, of the chestnuts and stones of the graveyard, with the microscopic landscape of the cell, of nucleus, of genes, cytoplasm, vacuole and wall, the landscape of science. Although on different scales, these landscapes are all part of our world.

Friday 13th February

ON THE PURIFICATION OF THALE-CRESS DNA

This morning, over breakfast, Alice asked me how we work with DNA. She is already developing the idea that separation, splitting of wholes, is a route towards understanding. That to study the DNA of plants it needs to be isolated, purified away from all the other things that make a plant. Whilst describing the process to her I was reminded of the first time I purified DNA by macerating seedlings in liquid nitrogen, grinding them with mortar and pestle until I could push no longer with the ache in my arms. The seedlings were transformed into an olive-green powder as fine as silt. A mud-spring of boiling nitrogen bubbles. The subsequent incubation of the extract with enzymes to digest the proteins.

Then the moment of revelation. My first sight of DNA. In textbook language, this was 'the precipitation of DNA from an aqueous solution by addition of ethanol'. Like this it sounds unremarkable. But I was entranced. I saw such flickerings and stirrings. A shimmering as of light in the air on a hot day. Strands of DNA were precipitating at the interface, moving through from solution into the layer of alcohol above. Their movement causing transient fluctuations in the optical densities of the two fluid layers, bending the light and making it dance. Whilst gently mixing these fluids one with the other, I watched the gradual seeding of a greyish-white fluff-ball of tangled DNA. And my excitement mounted still higher the following day, when, in the subsequent phase of purification, lit by ultraviolet light and marked by a dye that shines in light of that wavelength, the DNA was visible again. Glowing orange-yellow in the dark, buoyant in a solution of graduated density, fluorescing, billowed, and lumpy as a cloud on a fine summer's day. The glowing was of course a chemical, a physical, thing. But it provoked awe. There before my eyes was that wonderful dread stuff of life. Emitting light like Christ in some Northern Renaissance painting of the Nativity at night.

DNA has an invisible molecular structure. It also has an informational structure, is organised into genes. Of course the first geneticists could not see genes. They inferred the existence of the gene from studies of the nature of inheritance. But more recently we've come to know that a gene is a segment of DNA, and that the DNA within the gene is in the form of a linear sequence, a code. When a gene is active, this code is copied on to a second molecule, a molecule called messenger RNA (mRNA). The code is then read off the mRNA and used to construct a third type of molecule, a molecule that also has sequence, a molecule known as a protein. Proteins are the active components of cells, the things

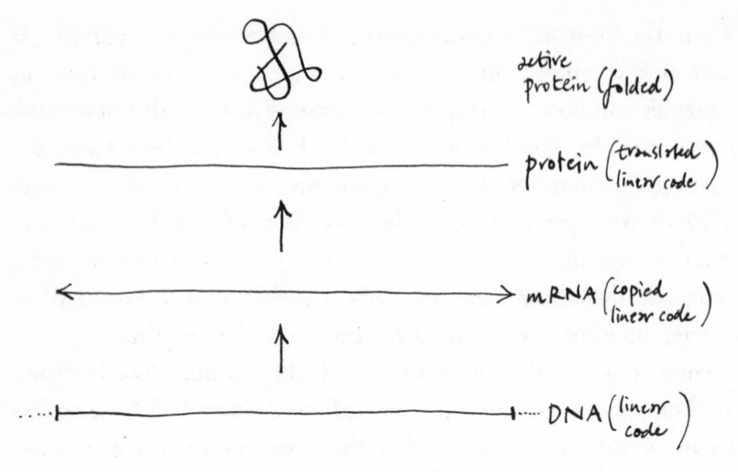

How DNA is copied into mRNA, which is translated into protein, which folds to make the final active protein structure.

that make life work. To do this they adopt a three-dimensional shape by the folding and wrapping of the linear sequence around itself. So genes (DNA) make RNA that makes protein.

All this I attempted to explain to Alice today. But breakfast was too hurried a time. And I had the sense that the words themselves – *DNA*, *mRNA*, *protein* – embody such condensation of meaning that they present formidable barriers to the exchange of understanding. Yet they are common currency within the scientific description of the process of life.

Saturday 14th February

A still, mild day. The sky plugged by a flat layer of cloud. Wheatfen a spectrum of browns from straw to rust. Water lower than when I was last here. The comprehensive landscape of the fen. Completed by its geography, by climate, by the organisms that live within it.

It's the same with the thale-cress plant and the grave on which it grows. The thalecress's present state is the product of an interaction between interior (genes) and exterior (the environment, the outside world). During the winter, the thale-cress detected the cooling of the air and activated genes that encode proteins that protect its cells from the cold. How all of these different proteins work is unknown. But something *is* known about one of them, a type of protein known as a transcription factor. Transcription factors are remarkable proteins. Encoded by a gene, they in turn control the activity of other genes.

Here I am, using another word that could benefit from unwrapping: *encode*, *encodes*, *encoded*. And I flinch a little at its starkness. Yet I can think of no better one, of no word that more accurately encapsulates the capacity of the information of the gene to be translated into the structure of the protein. It is presumably for this reason that *encode* is a common word in the genetic vocabulary.

As I began to describe yesterday, genes are built from segments of DNA. There is the segment that encodes protein (the segment from which mRNA is read). Ahead of this is another segment known as the promoter. Binding of a transcription factor to a promoter can cause the copying of the protein-encoding segment into mRNA, and hence the subsequent translation of mRNA into protein. Different genes have specific DNA sequences within their promoters, and these different sequences are recognised by specific transcription-factor proteins.

To return to the thale-cress. The falling temperatures of November activated a gene encoding a transcription factor known as CBF. Activation of that gene resulted in the production of an mRNA (read from its protein-encoding segment) and the subsequent formation of a folded CBF protein. CBF then bound to the promoters of further genes (1, 2, 3, etc., as in the sketch) and activated them. These further genes encode proteins that can

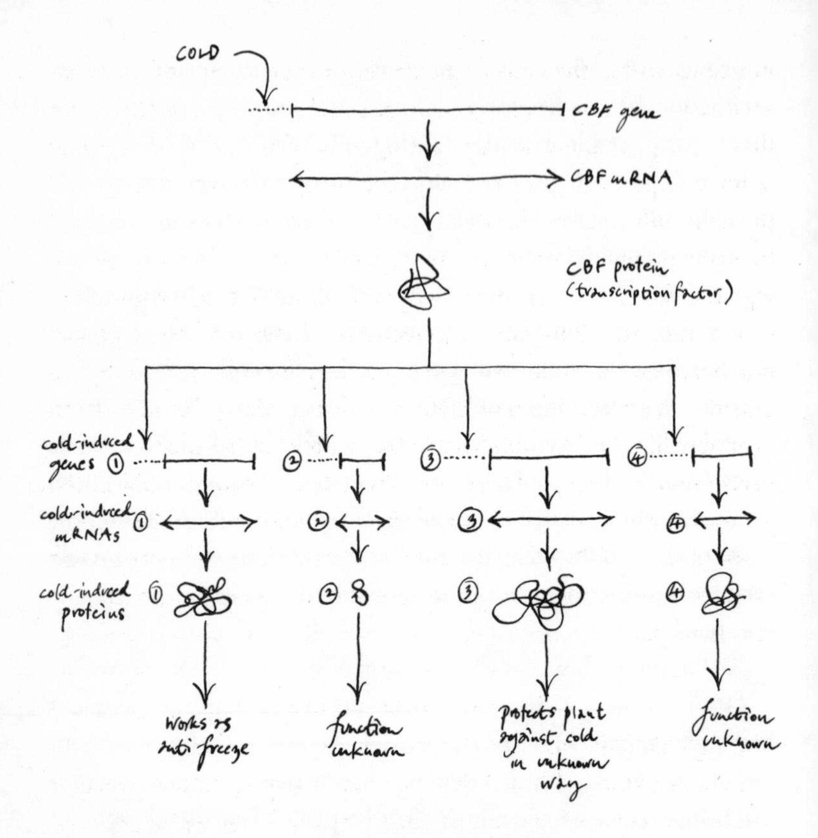

A cold-induced cascade of gene activation. Cold causes activation of the promoter of the *CBF* gene (dotted line): the production of *CBF* mRNA which is then translated to generate the active CBF protein. In turn, CBF binds to the promoters of genes 1, 2, 3, etc., and activates them. The protein products of these CBF-activated genes have different functions in the response of the thale-cress plant to cold.

protect the cells of thale-cress plants from the damaging effects of increasing cold. The whole process is a cascade of events that connects an initial stimulus to subsequent responses. The second wave of cold-induced genes encode proteins that perform specific functions within the cells of plants. For instance, one of them has anti-freeze activity. It changes the physical properties of the water that is so prominent a part of the cells of plants. The freezing point is depressed, thus reducing the danger that the contents of the cells will freeze. This ability of the plant to sense and respond to environmental change, an ability conferred upon it by its genes, was the factor that enabled the plant to survive the winter.

For me, this illustrates connectivity. Demonstrates a crucial link between the molecular world inside plant cells and the world outside. We often think of DNA as a distant thing. Remote from everyday life. Only visible within the context of 'science'. But this perception is wrong. There are direct links between DNA, the genes that are made from it, and the world around us. The world changes the activity of genes, gene activities change the properties of plants, and plants change the nature of the world. This circle is one amongst many that define the world in which we live.

Monday 16th February

The first daffodil open in the garden. Is this early?

Late last Friday I heard that our manuscript (see *6th February*) had been accepted for publication. Provisionally accepted, that is, with a request to make alterations and deal with questions raised by the referees. I'm very excited – and relieved as well. Actually the journal editor and referees dealt with this paper in astonishingly quick time. And the points raised by the referees will be relatively easy to deal with. I'm confident that 'final acceptance' is not far off. Although you never can be certain.

Thursday 19th February

Bright. Cool morning. Blossom trees beginning to sprout flowers. The first cream-yellow primrose in the garden. As the day progressed, a chill wind blew in from the east.

Alice and Jack have been asking why I go to Surlingham so often, why I keep going on about Wheatfen and St Mary's churchyard. So after school I scooped them up, together with their friend Tess, and drove them there.

First, to Wheatfen, where the wind was relentless, steady and penetrating. Alice was soon too cold. She became inward-looking, concentrated on her discomfort, unable to relish the kind of thing that normally kindles sparks in her. But tomboy Tess, with Jack in tow, was activated. They ran along paths, climbed trees with clumsy exuberance.

Then to St Mary's. I showed them the thale-cress plant. But they weren't much interested. 'Very nice, Dad' from Alice, still grumpy with cold, hardly glancing at the plant in her hurry to shelter from the wind in the church. Jack didn't see the point either: 'It's just a weed, we've got plenty of them in the garden at home.'

We went into the church, where they were briefly entertained by an eagle lectern, wooden, with an owl beneath. But soon they were on to other things. Clapping and shouting to hear the echo, hopping along the aisle, racing from west to east. I hurried them out, uneasy at the disturbance of the peace, bundled them into the car, and took them home. But I'm glad they've seen it all.

Monday 23rd February

Awoke to bright light penetrating the gaps between the curtains. Reflected in off a thin layer of snow. The lawn flat white, with clumps of tall grass poking through.

Took the bus to work because of the snow. Front seat, top of a double-decker. Spread out before me a wide horizon. On one side the purple-lead of an approaching snow-cloud. I thought about the fragility of life. The discomforts (the cold, the wet feet) of this weather seem to highlight the fit of life with the earth. How miraculous it all seems: there is a range of climatic variation within the atmosphere that life can tolerate. But the rest of the universe, out there in the blue sky, is extreme, beyond that range. An extremity that would instantly evaporate or snuff us out were we exposed to it. The earth is unique.

Wednesday 25th February – Ash Wednesday

The time for Lenten fasts. For austerity, frugality. I'll try to adopt such cleansing to my thought. Perhaps thinking sparsely and sinuously will reveal the nature of the problem in black and white.

Friday 27th February

THE MERISTEM – THE DIVISION AND EXPANSION OF CELLS

Another fall of snow in the night. This morning it is still cold – the snow remains, is not melting. I've been sitting at my study window watching the light. Earlier it was brittle-radiant. But in the last few minutes it has dimmed. Now specks of snow are starting to fall. Whilst watching, I've been thinking again about the thale-cress. That plant in St Mary's. Mind buzzing with excitement. But there is too much snow for me to go to Surlingham today. In any case the plant will be completely buried. I'll use imagination and memory to fuse a description of it.

The centre of the thale-cress's rosette of leaves is the source of all the rest of it. Each of the thousands of cells from which the rosette is built can be traced back through a line of descent to a founding cell in the centre. To a unique cluster of special cells, a structure known as the meristem, at the very centre of the centre of the disk. Everything seen as the rosette-leaf disk originates from that unseen meristem.

Actually, the meristem is at the stem tip. As in all plants, the thale-cress shoot consists of leaves arranged around a stem.

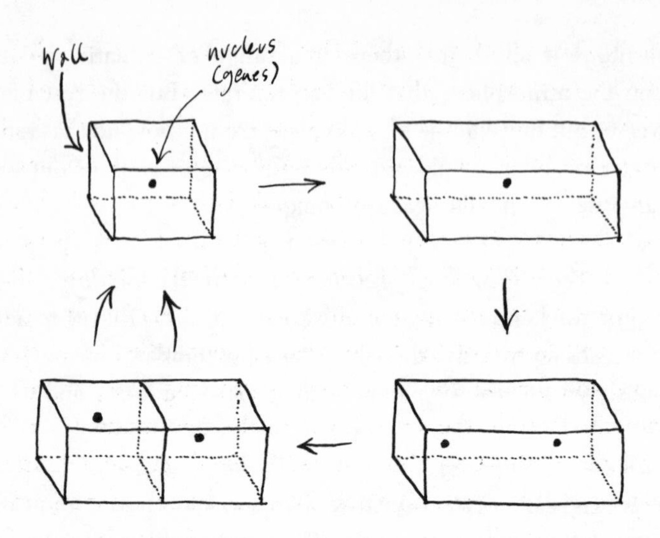

How one cell makes two cells. The cell first expands, then replicates the genes in the nucleus, then separates the two sets of genes into separate nuclei, then builds a new cell wall to divide the old cell into two. Each of the new cells can then begin the process again.

But at this stage of the thale-cress's existence, that stem is squashed to a length that makes it hard to see, hidden by its shortness and by the leaves packed around it. During growth, cells are generated in the meristem, then flow downwards to form the main body of stem and leaves.

The growth of plants depends on the division and expansion of cells. These processes can be visualised. First, imagine a cell. An abstraction of the real thing. In the shape of a cube. With walls, an inside, and a nucleus containing a complete set of genes. An imagined typical cell (not that there is such a thing). The cell gets longer, the cube becoming an oblong. Then it copies its genes. Then it pauses for a while. Then the two sets of genes separate into the two different ends of the cell. Then the cell builds a new wall, splitting itself into two. There are now two new cells, each with its own set of genes, where before there was one. It is division and multiplication at the same time. Two cells made from one. And then these two cells themselves start to expand, beginning the whole cycle again. This is how growth works at a cellular level. The growth that we actually see is the product of the cycles of division and expansion of many different cells.

Furthermore, the proliferation of cells enables the shoots, roots, and leaves of the thale-cress to adopt their characteristic shapes and forms. The thale-cress is growing now, at this very moment. Despite the snow, even in the few minutes I've been writing and thinking about it there has been an infinitesimal progress in that essential mathematics. Each cell of the meristem is now a little closer to becoming two cells than it was when I began to think of it. With spring on the way, as the air warms, so the rate of proliferation will accelerate. And I find that there's something comforting about the idea that the rate at which the cells in that thale-cress meristem proliferate is dependent on the angle of tilt of the earth with respect to the sun.

Why comforting? What do I mean by this? As soon as I've written it I'm aware that there's more to it. But that more is hard to grasp. It is fleeting, unstable, difficult to express. Fades as soon as it's captured, slips through the mind. But the declaration of this sense of comfort is at least an identification that this record of the growth of the thale-cress plant is, for me at least, more than a simple, detached observation of life-cycle progression.

Saturday 28th February

The whole country is gripped by cold northerly air. There are blizzards in Scotland, and temperatures there fell to –8 degrees centigrade last night. Here in Norfolk, although yesterday's snow was partially thawed by the sun, we had yet another heavy fall in the night. The view from my window is of hedges and bushes all bobbly and lumpy, the shape of leaves exaggerated.

Again, it will be difficult to get to St Mary's. Nor would there be any point to it. So back to contemplation. To seeing the structure of the meristem of the thale-cress in the mind's eye.

Again, the vision is an abstract one. It focuses on central things, omits the peripheral. It sees the meristem as a tiny ball of several hundred cells, a sphere of about 150 micrometres in diameter. The cells in the meristem are smaller than many of the cells in the rest of the plant, because meristematic cells divide after only a relatively short phase of expansion. Inside their walls, they are filled with a gel made of proteins and other macromolecules suspended in water (this gel is known as the cytoplasm), and a nucleus (the structure that contains the genes). The meristematic ball of cells is at the apex of a dome, and that dome forms the tip

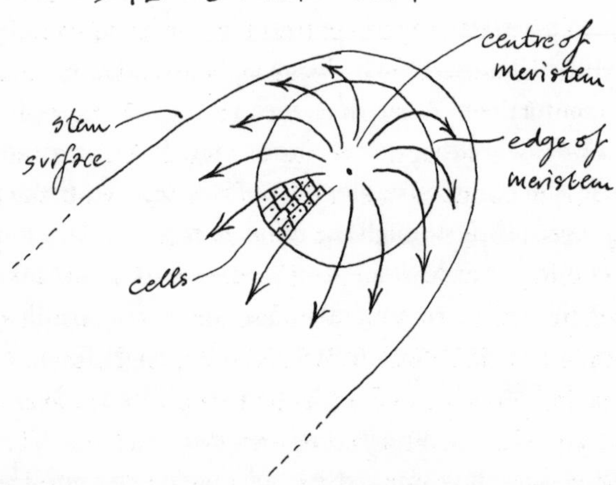

How cells from the meristem make the stem.
Cells are made in the meristem (*the ball in the diagram*),
then flow out in the directions shown to make the stem.

of the plant's stem. And although I envisage it as a sphere within a dome, and have drawn it in this way there is in reality no sharp line that divides the cells of the sphere from the cells of the dome. The sphere is in some ways a conceptual device that enhances the definition of the part of the plant that is principally concerned with the making of the cells from which the rest of it is made.

It's important to remember that the cells of the meristem are not static. On the contrary, they're in constant motion. The meristem is the source of a stream of cells that flows out of it to become the body of the plant, to make the stem and the leaves. Hence the flow arrows in the sketch. Most of the cells that comprise the meristem are but a fleeting part of it, are made there and then pass on to make the rest of the plant.

There is an internal genetic mechanism that enables the structure of the meristem to be maintained. This structure is dependent on the rate at which the cells that comprise it are generated. Cells at the very centre of the meristematic ball stay relatively small and divide only infrequently. Upon division, a single daughter cell is pushed further out from the centre, towards the ball's surface. The further a cell is from the centre of the ball, the more frequently it will divide, with subsequent daughter cells being pushed out even faster.

On the surface, cells divide and expand with relative rapidity. These relative rates of cell proliferation are controlled by a protein called WUSCHEL. WUSCHEL (named after a German word meaning 'disorganised') was first identified because mutant plants lacking it have a disorganised structure. In fact, WUSCHEL makes the cells at the centre of the ball divide at rates sufficient to generate the cells that are eventually used to make stems and leaves. In plants lacking WUSCHEL, the cells at the centre of the ball don't divide, and the cell source dries up. In plants having too

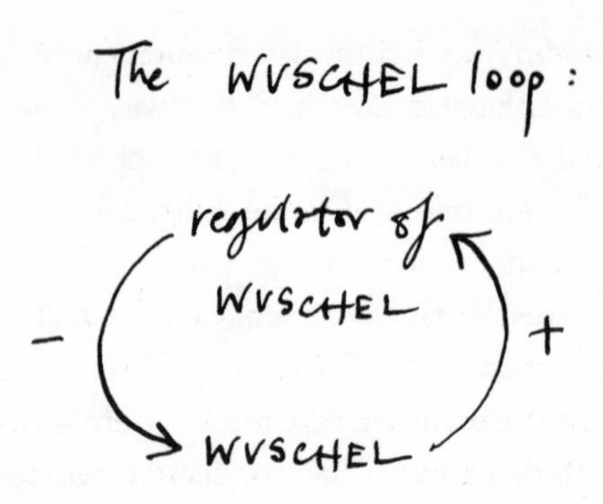

How WUSCHEL works. WUSCHEL promotes the proliferation of cells in the centre of the meristem. But WUSCHEL also promotes the production of 'regulator of WUSCHEL', that itself inhibits the action of WUSCHEL. This is a 'negative feedback' homeostatic loop.

much WUSCHEL, the cells at the centre of the ball proliferate excessively. Furthermore, the levels of WUSCHEL in normal plants are 'feedback'-regulated by a second protein. There is a gene that encodes that second protein. This protein (we will call it 'regulator of WUSCHEL') inhibits the production of WUSCHEL by inhibiting the expression of the first gene, the gene that encodes WUSCHEL itself. In turn, WUSCHEL activates the expression of that second gene. This kind of controlling loop is a 'homeostatic' loop – something that secures the constancy of things, a common refrain in biology.

The WUSCHEL loop is a small part of the mechanism that ensures that the meristem maintains a constant structure despite the fact that the cells from which it is built are flowing through it. One can picture how some of the genes of the cells of the meristem make a pattern – a map – of signals. Other genes control the activity of the cells of the meristem with respect to that map. It is a map that is being made at the same time as it is being read, and the reading affects the making and the making affects the reading in loops of subtle complexity that we are only just beginning to understand.

Sunday 29th February

HOW LEAVES BEGIN

Yet more snow last night and today. It's falling now. The consistency and texture of the flakes are different from yesterday – larger, more fluff than hard. Then they were sharper – more punctate. Now garden and sky are connected, washed in the one white. The snow lying on twig and branch has a shadowing quality; it marks the surfaces, brings them into relief, separates the tops from the bottoms of branches, makes the trees sing with dimensionality as if they were in an etching.

I expect that the thale-cress plant will be lying beneath a layer of snow. So this will be a third day of visualising its meristem. Yesterday's image of ball and dome is the foundation to today's picture. And today's picture is to be more complete, will depict the formation of the leaves and explain their positioning with respect to one another.

To begin with, an outline description of the way in which the leaves are arranged on the plant. Each leaf is attached to the central stem by its stalk. In total, these attachments make the disk of the leaf rosette. And that rosette is actually in the form of a flattened helix. The position of each leaf relates to that of the previous one by a turn of the helix, that helix being entwined around the stem from top to bottom. The geometry is exquisite. The angle described by a leaf, the leaf that follows it, and the centre of the stem is always 137 degrees. And this 137-degree angle is set in the meristem.

A leaf begins when a small group of cells on the very flank of the meristem dome starts to divide more rapidly than their neighbouring cells. This selected group begins to divide outwards,

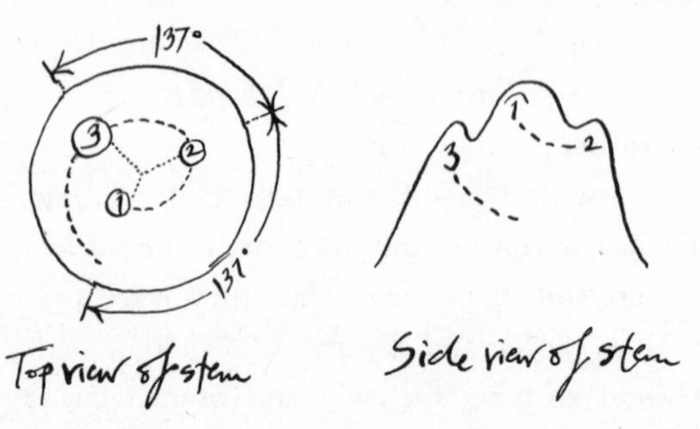

How leaves begin. Bumps 1, 2 and 3 represent succesive leaf primordia forming in a spiral around the meristem and stem tip.

making a bulge on the side of the dome. That bulge is the first sign of what will become a leaf and contains the founding cells from which it is to be built. Soon, another bulge will form, and then another. The site at which each bulge forms relates to the position of the previous bulge; there is a constant 137-degree angle described by successive bulges and the centre of the meristem. And because these bulges eventually become leaves, the microscopically visible distribution of successive bulges around the meristem generates the helical pattern with which the fully formed leaves are arranged around the stem. The unseen creates the seen. Exactly how the 137-degree angle is determined remains one of the most tantalising unsolved problems in plant biology.

So now I have it. A consolidated mental image of the thale-cress shoot. With luck the meristem will not be seriously damaged by today's snow. As the spring progresses so the rate of cell proliferation will increase. The first leaves are already there, dormant bulges waiting for warmth. As the temperature rises, so the speed with which the plant adds leaves to its spiral will accelerate.

There's something pleasing about the mathematics of all this. That the entire structure of the thale-cress rosette can be abstracted down to a single angle, the angle of arc separating successive bulges on the surface of the meristem.

Later – it is dark now and the snow is falling again. Driven obliquely by the wind against the light in the street. I imagine it falling all across the land – from here in Norwich, to the thale-cress in St Mary's, to Wheatfen, to Yarmouth, and to the great grey North Sea beyond.

This natural-history project is fun. I'm enjoying it. But still I haven't solved the problem with which I began.

MARCH

Monday 1st March

THE ANATOMY OF CELLS – DISTINGUISHING LIFE FROM NON-LIFE

A STRONG sense of seasonal progression this morning. Bright light penetrating past the edge of the bedroom curtains *before* I get up to make our early-morning tea. Cycling to work in hard cold. Frost, frozen snow crackling under the tyres, shattered ice in puddles. At a guess, four or five degrees below freezing. But I have energy this morning.

And yesterday, after the last heavy fall of snow, the sun began to melt it. Wet slush fell from the trees in showers. Standing in the garden, the sound of it all was wonderful, the tapping and clicking of falling drops, the rushing sound of slipping accretions of snow, the running and bubbling of water falling in the drains from the roof. Like a spring. A mountain stream heard through a babble of birdsong. I was aware of a sense of resurrection. But I do veer between the sublime view and that of detached objectivity.

This week's work: to make final corrections to our revised paper, then return it to the journal editor. Tomorrow a meeting with a group from another UK research institute. Friday to Warwick. In between these meetings, the running of my research team. Sometime I must start writing a grant proposal.

To return to the imagination. Just for a few minutes before the real work of the day begins. For fun. Last night I brought home a thale-cress plant from the lab for Alice to look at with the microscope. She was entranced by what she saw. By the landscape of the leaf surface, by seeing the cells of that outside face and the pores that admit entrance to the inside. We imagined what she might see were she to shrink, Alice-in-Wonderland-like, to a size that would let her through one of those pores, into the chamber beyond. And now I am similarly imagining the cells of such a chamber. From the lower side of a leaf of the thale-cress in St Mary's. I'll go in layer by layer, the sketch will serve as a guide.

The surface consists of a layer of epidermal cells, the outer skin of the plant. Scattered throughout the epidermal cells are pairs of cells known as guard cells. These form gates to the pores, gaps in the epidermis through which the outside air connects with the air-chamber. The guard cells shrink or swell in response to signals from within the plant, thus opening or closing the gates. Today, because of the searing cold, those gates will be closed. But in warmer weather those gates will be open, with air flowing in and water vapour flowing out of the chamber. The walls of the

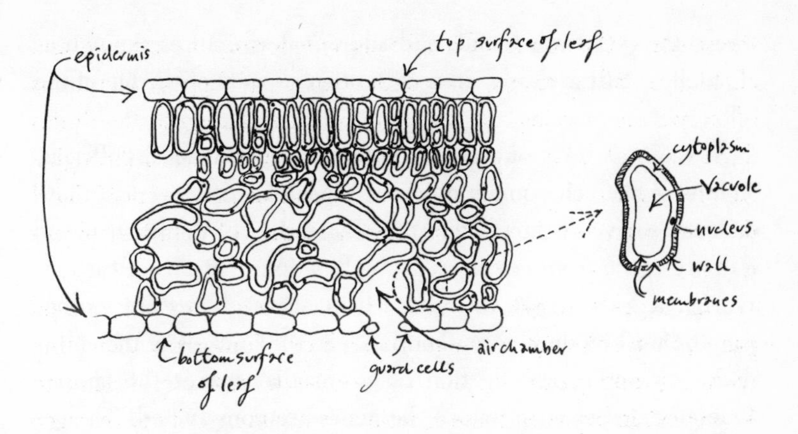

A cross-section of a leaf, showing the guard cells of a lower-surface pore, the air-chamber it serves, and a close-up of one of the cells that forms the wall of the chamber.

chamber consist of the surfaces of internal cells of the leaf. These cells absorb carbon-dioxide molecules from the air that enters the chamber, then use this carbon dioxide in the process of photosynthesis (something to which I will return later). The carbon dioxide begins its journey into the cells by passing through the cell wall, the outer face of those cells. A wall made from molecular fibres woven together. A fabric, like cloth or paper.

The cells themselves consist mainly of water. At the centre of each cell is a droplet of water held within a thin membrane, the vacuole. Surrounding the water droplet is a thin layer of cytoplasm. So the structure of the cell can be thought of layer by layer: first there is the wall, a protective paper bag within which the cell is sealed. Inside the wall, wrapped in a membrane, is the cytoplasm. Inside the cytoplasm, held in another membrane, is the central vacuole. Although only a small part of the whole, it is the cytoplasm that is the meat of the cell. And the cytoplasm is itself subdivided: it contains the nucleus, the place where the genes are found, and additional structures (known as organelles) that are associated with other aspects of the life of the cell.

At the next level of abstraction there are molecules. Different regions of the cell contain different types of molecule, and these molecules have distinct properties. For instance, the fibres in the wall are made from cellulose, a combination of atoms of carbon, hydrogen, and oxygen in a long chain. These fibres are strong, give the wall its durability. The DNA in the nucleus of the cell is also a molecule, a collection of atoms that makes the famous double-helical chain of which the genes are constructed. The gel of the cytoplasm is a thick soup of other molecules: proteins, carbohydrates, lipids, all suspended in water. The molecular constitution of the cytoplasm is very similar to that of the cells of our own bodies.

Life is in layers. We can see it at different levels of abstraction. On different scales. There are the molecules, the proteins, the carbohydrates, the cellulose, and the DNA. There are the internal compartments of the cells, the walls, the cytoplasm, the vacuole, the nucleus. There are the cells themselves, the thale-cress in St Mary's churchyard, the chestnuts surrounding it, the nearby fen, the more distant North Sea. And a question forms in my mind. An unexpected question. Is the whole of the cell alive, or is only a part of it?

This is a question I haven't previously considered. A simple question of such obvious importance that I can't imagine why I've never thought of it before. What is alive and what is not? Where is the boundary between life and the absence of it?

A thale-cress plant is alive. The cells of that plant are alive. But at the very centre of most of those cells is a vacuole. Is the water inside that vacuole alive? Or the walls that surround the cells, are they alive? I think that the cytoplasm of the cells is alive. It moves, it respires, it metabolises. Yet it is made of molecules which are, it seems to me, not. The genes? Are the genes alive?

And now a further image arises in my mind. Of a petal detaching from a rose. The petal falls to the ground. Descends, twirls through air to earth. Then begins to decay. Crimson fades into brown. The petal shrivels and twists. Dark veins, a sepia background. The process of disintegration. The cells breaking down into the molecules from which they're constructed. Those molecules leaching into the water of the soil. Further disordering of the molecules into their constituent atoms and ions.

Finally, the petal disappears. And although it seems as if it had never existed, those once-incorporated ions, atoms, and molecules are now distributed in the earth and air. Then we move on a few years. To a time when a few of these same atoms have once again become part of a living thing. Of a grass stem picked by a child so she can suck in its sweetness. Atoms, once of the petal, then returned to earth, then of the grass, are now of the child.

Wednesday 3rd March

I was this morning putting a few final corrections into the revised version of our paper . . . changing 'which's to 'that's and so on . . . when one of my research team asked me to come and look at something.

So I went to look. We discussed the result, evaluated it critically, debated its possible meaning, etc. But what was in my mind was wonder. Wonder that we could see what we were seeing, let alone be considering its potential significance.

The experiment in question had enabled us to explore the arrangement of the DNA in a particular gene. And what we had seen was exciting. Suggested that we had got what we wanted. We'd hoped to find a plant carrying a mutation in this gene. It seemed that, after months of searching, we'd at last found what we'd been looking for.

But how to communicate this excitement? The reasons underlying it, the context, are key. Yet it's hard to achieve a picture having the full depth and resonance of context, because the language with which that picture is painted is not common. This is the problem of specialisation. It divides us. We're split into individual cells that talk about the world in different languages. And although we strive to comprehend our world, we're failing to generate a collective vision.

Thursday 4th March

Mild. Grey light. Homogeneous mist at ground level merging into homogeneous cloud.

Our paper is now fully accepted for publication! An amazingly fast response from editor and journal. I submitted the revised version late yesterday afternoon. First thing this morning there was a 'paper accepted' message in my e-mail. And I've been thinking some more about the cells of the thale-cress plant. Earlier I sketched a picture of how the cells are separated from one another by the walls that surround them. But this is only partially true. The separation is not complete.

It is, to an extent, a question of how the cells are seen. They are indeed places apart. Islands, distinct from one another, separated from the rest of the world, contained within membrane and wall.

But the cells are a community. They communicate with one another in a physical sense. There are tiny threads of cytoplasm that penetrate the walls and link each cell to its neighbours. The cells communicate with one another via a network of such threads. Substances that carry messages pass along them. In addition, the membranes and walls are selectively permeable.

They regulate the traffic of particular substances from the cytoplasm of one cell to the cytoplasm of the next (via a route quite independent of the connecting threads of cytoplasm).

So the cell-islands belong to archipelagos, are at once separate and connected.

Saturday 6th March

THE DEVELOPMENT OF LEAVES

To Wheatfen broad and woods with Alice.

Slight chill in the breeze but actually quite mild. Sky grey but some heterogeneity to it. Sun-disk fleetingly visible through cloud.

Out into the expanse of the fen on a grassy path to a seat beneath a willow. The distant sound of waterfowl. The fen a lovely collage of browns. Sandy-cream reed stems with chocolate tops, compact chestnut bulrushes. The reed stalks stacked and layered in patches of parallel lines. And at the base of the reeds – as if with the joy and pain of birth – the first green spikes of the year. Flattened leaf-pyramids, with tips and tapering straightness below. But too chilly to sit for long. Alice tugging at me to move on. The cold beginning to clamp my fingers as I write.

When we get into the wood the trees are in silhouettes against the light. So much stick and twig. Reticulate. The tapering relationship between trunk and branch and twig so clear in the absence of leaves. Honeysuckle at the base of an oak sprouting green. Daffodils and snowdrops near the car-park.

Then to St Mary's. The sky slate-grey by now, with a brighter rim on the horizon. A shower. The churchyard seems shut within its wall of dour chestnuts. A sense of the dismal enhanced by a bunch of frost-nipped blue pansies lying pathetically on one of the graves. And as we approach the grave on which the thale-cress grows, I can see that there has been a change. An alteration in the texture and colour of the earth. More grit and less greenery than before. The soil more lumpy. With alarm I realise that someone has been tending the grave.

Of course I knew from the start that this project has risk attached to it. That at any moment something might halt the progress of the life I was recording. But it's a false alarm. To my relief, I find that although the rest of the grave is now bare earth and gravel, the thale-cress and its immediate neighbours are unscathed. Saved by being tight in the corner of the graveside wall, by the weeding being unfinished. I wonder why the job was left incomplete?

Since last I was here, the thale-cress has been adding to the spiral that makes the rosette. Two more leaves can be seen. Cells have moved from the meristem into the bulges. The cells in those bulges have themselves been dividing and expanding in different directions. The wonderful thing about this increase is that it is not a chaotic proliferation. No. The leaves grow from the bulges by a co-ordinated series of cell divisions and expansions that collectively result in an organ having a final identity: a recognisable leaf, a more or less flat plane constructed from several layers of cells.

If the generation of a leaf is not totally chaotic, neither is it rigidly defined. It straddles the boundary that divides the random and the determined. This growth can, on one simplified level, be expressed as a mathematical model. The bulge on the side of the meristem consists, at a particular moment following its initial establishment, of x cells. Each of these cells will then proliferate through several generations, say y generations, to make the total number of cells of the leaf: $x \times 2^y$. The 2 in this equation expresses how each time a cell divides, two cells are made. Because plant cells are fixed to one another by their walls, the descendents of each of the original x cells tend to remain in a clump as they proliferate. So the leaf is made of x patches of 2^y

cells, each patch a clump derived by descent from one of the original cells of the bulge. Each of the original cells can be thought of as having a future path of divisions and elongations that it will follow. A path that branches at each cell division, on each occasion when one cell becomes two. The sum of these paths is the construction of the leaf.

Now let's add geometry to our algebra. Let's imagine that each cell is a box with six sides. If we restrict ourselves to perpendicular divisions, there are three different ways of dividing the box into two boxes: top to bottom (1), top to bottom again (2), and side to side (3). So there are three different possible planes of division for the cell.

When a cell divides, it chooses one of these planes. And that choice affects the final shape of the clump of cells that develops from it. Further shaping of the clump is achieved by influencing the relative directions with which cells subsequently choose to expand (for instance, do they expand one wall in a certain direction and not expand another?). The shaping of the clumps adds up to the shaping of the leaf.

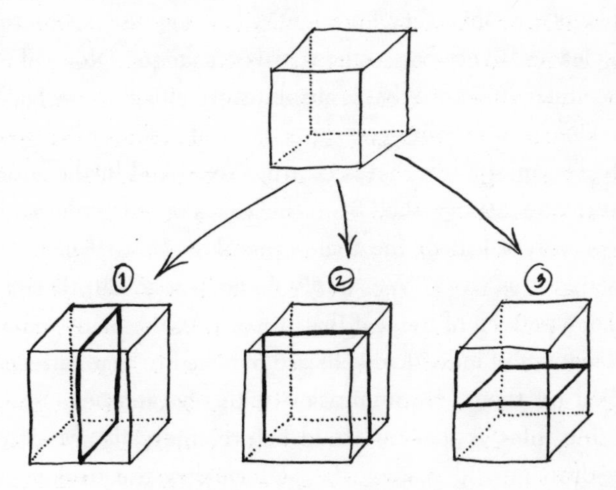

Alternative ways of dividing a box.

Now here's a remarkable thing. It's clearly a highly complex matter to organise the divisions and elongations of the *x* cells to end up with a leaf of a certain shape and size. Perhaps the simplest solution to a problem of this complexity would be to devise a precisely defined path for each of the *x* cells (divide in the one plane, then elongate so many micrometres in the same direction, then divide in the opposite plane, etc.) so that the combined paths for each cell added up to a leaf. Not only is this the simplest solution, we already know that it happens during the growth of other organisms. For instance, some species of nematode worm organise the fates of cells during development in a similarly precise manner.

But leaves don't grow like this. Each time a leaf grows, the cells adopt fate-paths that we can think of more as tendencies than as precisely defined routes. These tendencies are flexibile, not rigidly prescriptive.

Each of the leaves of the thale-cress plant has grown by a distinct series of cell divisions and expansions. This is just a single plant. Throughout the world there are millions of other thale-cress plants, at this very moment all doing the same thing, growing leaves. Every one of these leaves is unique. None of them share a cellular structure that is absolutely identical to the leaf that I am looking at now. But in all cases the final result of leaf growth is much the same: a leaf that is clearly a leaf. And furthermore, a leaf that can be distinguished from the leaves of other species, that is recognisably a leaf of the thale-cress *Arabidopsis thaliana*.

How does this work? We simply do not know. But there must be a map, a pattern of the leaf that the cells of the plant generate. The behaviours of individual cells are matched to that pattern and controlled by their communication with one another. And the remarkable thing is that, as I said before, the cells are simultaneously drawing and reading the pattern they use to determine their behaviour.

Another shower has begun. Alice and I are both feeling cold, the wetness soaking through our clothes. On the way home she asks what I was thinking about, standing there staring so long at that weed. And I talk to her a little about the cells and the growth of the leaf. She is interested, likes to see things in her mind. And is astonished that each leaf, whilst outwardly so similar, is the product of a unique series of cell divisions and expansions. So unique that the world has not seen it before, and never will again. It reminds her of something she has recently read, of how snowflakes, although each one generated by the same set of rules, are all individually unique.

Monday 8th March

Yesterday, in raw, damp cold, walked through Chapelfield Gardens. There's an avenue of trees on one side, beneath them a spread of crocuses. Plain white and yellow ones, others with longitudinal purple stripes. Shining amongst the mud. They excite me: despite their apparent fragility they have survived the recent frosts.

Later, during dinner, I suddenly found myself thinking about Darwin, and *The Origin of Species*. About the great idea of evolution via natural selection. It's a defining concept in modern biology. A crucial insight into the nature of life. Its subject is how life is divided into species and how those different species came to exist. Yet, balancing this, Darwin's vision also shows that the different species have more in common, more uniting them, than they do that divides them.

This morning the cold is returning in wind from the east. There may be snow showers tonight. And I'm out of sorts. Back stiff, and there's an ache at the base of my spine. I'm unsettled, thinking that I'm getting old, not liking the thought.

And I'm also suddenly uneasy and confused about the recent

race to publish our paper. Should we all have been communicating more? Keeping each other in the picture when planning experiments etc.?

Yes, well. Perhaps we should. But it's so difficult knowing where to draw the line. What the rules that define that 'should' actually are. The game is a competitive one, no question. The rewards are for being first. It follows that if there's a risk that talking will compromise primacy, talking tends not to happen. In any case it's so difficult to plan. The path of science is never clear. A line of investigation that might not seem interesting today could become of crucial importance tomorrow.

But I don't like all this. It tarnishes the beauty. I wish it were otherwise. I'm uncomfortable with the feeling that my actions, irrespective of right or wrong, might be hurting someone else. This aspect of what we do creates anxiety, tightens my stomach. But today I will act. Project confidence. Be as noisy as some drugged-up hero going into battle. Suppress the anxiety. In the back of my mind, though, there is the thought that heroes create tragedies.

There is of course an opposing view. That competition keeps you honest and busy. Honest because you can't afford to publish things that are wrong when you know that your experiments are probably being duplicated by someone else. Busy because you need to remain ahead of the game.

Tuesday 9th March

HOW GENES MAKE THE TOP AND BOTTOM HALVES OF LEAVES

A bitter, cold, and penetrating wind. Sleet pings against the window-pane. I wish it would get warmer. Funny how winter and spring, cold and warmth, dance around each other like weather-men and -women coming in and out of a weather-house.

To continue with the development and structure of leaves. As the cells of the bulge on the flank of the meristem make the structure that has the identity of a leaf, so that leaf establishes within itself areas which have their own regional identities. Leaves have an upper surface that faces the sun and a lower surface that faces the earth. It's as though you could split the flatness of the leaf down its middle, like you can split a piece of shale along its long edge and make two new planes: a top half and a bottom half. This division has biological meaning. The top half of the leaf is more concerned with the harvesting of sunlight during photosynthesis, whereas the bottom half is more for gaseous exchange, for the absorption of oxygen and carbon dioxide from the air. These identities – top half, bottom half – are written in the genes. There are genes that encode proteins that tell the cells to become bottom half or top half. These proteins act early in the formation of the leaf, when it is still a bulge on the side of the meristem, and

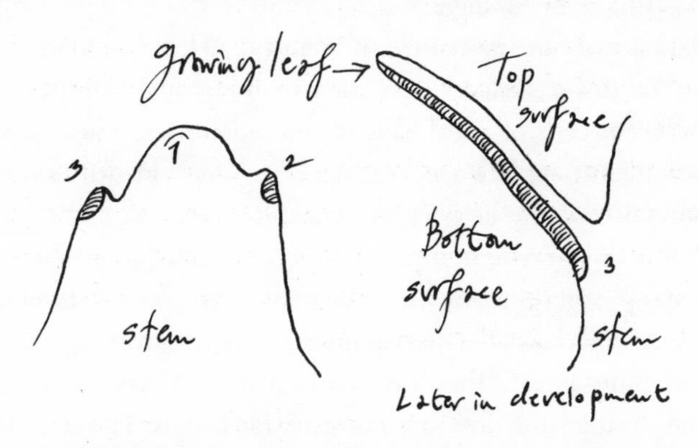

The location of the 'bottom-half' transcription factor in developing leaves. Shading shows where this transcription factor is located. *Above left*: in the bottom halves of primordial leaf bulges (2, 3). *Above right*: in leaf 3 after it has grown further. The transcription factor is restricted to the bottom half and is not found in the top half of the leaf.

divide that bulge into two territories. For example, a particular transcription-factor protein is found in the bottom half, but not the top half, of the developing leaf bulge. This transcription factor tells those cells containing it that they are part of the bottom half of the leaf, by activating the particular subset of genes that are needed to make 'bottom half', as opposed to 'top half' cells.

Wednesday 10th March

THE NATURE OF IDENTITY

To St Mary's again. Despite the penetrating cold of an east wind. The plant is still growing, seemingly unaffected by the recent disturbance of the soil nearby. Leaf growth has continued, rates alternately pushed and slowed by diurnal temperature variations. Faster in the warmth of day, declining in the cold of night. But today my fingers ached with the cold, and whatever growth there was will have been slight.

Thinking about the nature of identity. A leaf has identity, the upper and lower surfaces have their own regional identities, even the different cells of a leaf have distinct identities. These separate cellular identities, like the 'leaf' and 'regional' identities, are the consequence of the activity of genes.

Within the meristem, the place where the cell lineages begin, the cells are prototypes. Small, with dense cytoplasms, lacking vacuoles. As cells leave the meristem and enter the growing bulge that will become a leaf, they begin to enlarge. Vacuoles form and expand. At the same time, a transformation begins. The cells start to adopt one of several distinct identities. Differing cell types become apparent. These cell forms have resonant names: the vessel cells, xylem and phloem; the parenchyma cells; the cells of the spongy mesophyll; etc. The process of transformation, the assumption of these various cellular identities, continues as the leaf grows.

How does this miraculous transformation occur? On the

surface of the leaf, from its earliest primordial bulge stage, is the single-cell layer known as the epidermis. The plant's skin. A relatively uniform layer of cells that are very similar to one another in shape and size. As the epidermis expands along with the rest of the growing leaf, some of the epidermal cells assume a new identity. There are choices being made here. Most of the cells continue to develop as standard flat epidermal cells. But a few choose otherwise and commit themselves to becoming one of two different types: a guard cell (described previously) or a trichome.

Trichomes are hairs, scattered in a pattern somewhere between random and regular across the surface of the epidermis. It is the layer of trichomes that gives a leaf its velvety feel. Yet the trichome is an extreme identity for a cell to assume. A spiky branched thing with a tough cell wall that sticks up above the flatness of the epidermal plane. A huge transformation from the basic epidermal cell. And this transformation is caused by the action of genes. Amongst the 30,000 different genes that the plant contains, there is one that is crucial to the transformation of young epidermal cells into trichomes. This gene is named *GLABRA1*, from the Latin (glabrous) for smooth, bald, or hairless. *GLABRA1* encodes a protein: GLABRA1. Thale-cress plants that lack GLABRA1 have a smooth epidermis and are completely lacking in trichomes. This is because GLABRA1 is a transcription factor, a protein that acts on genes. It activates them (and perhaps inhibits other genes), creating the particular pattern of activity that transforms a cell into a trichome. In the absence of GLABRA1, this pattern of activity is not established, and the cell fails to become a trichome.

In fact, during the development of a leaf, GLABRA1 activity is progressively focused in those cells that are to become trichomes. When the leaf is at the bulge stage, *GLABRA1* mRNA is found in all cells of the epidermis. As the leaf develops further, so the mRNA's location becomes restricted. It disappears from some regions, remains in others. Where it remains, it is seen in patches of contiguous cells, then in smaller patches, and is finally focused in individual cells. It is these single cells that then become trichomes, in response to the mRNA they uniquely contain. How the progressive restriction of *GLABRA1* expression is achieved remains unknown. A fascinating phenomenon.

A note about genetic nomenclature. Genes are often first identified because something happens when they fail to work. For instance, the *GLABRA1* gene was so named because mutants in which that gene fails to work lack hairs. But this naming can seem counterintuitive. The function of the normal *GLABRA1* gene is actually the opposite of what that name might suggest. *GLABRA1* causes the formation of hairs. This paradox is found throughout the genetic nomenclature of thale-cress, applies to many of the other genes that will feature in these notes as the story of the plant develops. It sometimes helps to think of it in terms of positive and negative images. Like a photographic negative and its resultant print: both valid representations whilst fundamentally opposed.

Whilst on the subject of nomenclature, here is another convention. A gene name, like *GLABRA1*, is written in *italics*. The protein that gene encodes, like GLABRA1, is written in normal script. This convention promotes clarity, enables a precise distinction between a gene and the protein it encodes.

Thursday 11th March

HOW THE STRUCTURE OF LEAVES RELATES TO THEIR PHOTOSYNTHETIC FUNCTION

The diversity of cell types within the thale-cress plant is evocative of the diversity of life itself. And diversity relates to function. The leaf is constructed of different types of cell that are specialised to the performance of different tasks. Sub-tasks of the overall task of the leaf. To give some examples, the leaf contains a meshwork of vessel cells (known as xylem and phloem). These vessels form a connecting network throughout the plant, carrying water from roots and nutrients from cell to cell. In another example, the epidermis carries a waxy cuticle and contains pores that regulate the flow of water and gases into and out of the leaf. Again, the cells of the parenchyma and spongy mesophyll are especially suited to performing the reactions of photosynthesis (of which more in a minute).

Yet this diversity has an underlying unity. Each of the transformations of cellular identity, each of the different cell types that make up the leaf, are the result of the action of different subsets of the genes that each cell contains. And it seems likely that these different subsets of gene activity are the result of the co-ordination of gene expression by specific transcription factors that switch on or off the sets of genes required for the development of particular cell types.

So what is it all for? The question formed this morning whilst I looked up from the thale-cress and across the graves, the oblongs of light and shade, mounds of grass. We live in a universe where disorder rules. Where order disintegrates. Why, then, these complex, organised structures? These leaves that the plant places in a spiral round its stem? What are they for? And as I looked back at the leaves, I had the strange feeling that they were themselves reminding me of their purpose. Squatting at the edge of the grave, I could see that the younger leaves, the newest ones, were as if hovering in a plane almost parallel with the flat of the land. But also that that plane was not absolutely parallel but had a tilt to it. That the flatness of the leaves was angled so as to be face on to the clouded sun.

Angled to the sun to maximise photosynthesis. That chemistry of huge subtlety and baffling complexity. The chlorophyll in the leaves absorbing light energy from the sun, then using this energy to separate the atoms from which water molecules are constructed, atoms bonded together with enormous strength. The energy liberated by the splitting of the water being subsequently used to make sugars from carbon dioxide. The sugars then fuelling the growth of the plant. A miraculous process that feeds all plants, and all animals, all of us. As I write, remembering what I saw this morning, it transiently seems to me that my writing about photosynthesis, pencil scratching across the pages of my notebook in the fading evening light, is photosynthesis writing about itself.

The structure of leaves facilitates photosynthesis. They have been shaped, optimised to this purpose during evolution. That's what has driven their organisation. There are three requirements for photosynthesis to occur: sunlight, water, and carbon dioxide. The thale-cress leaves display themselves as flat planes perpendicular to the flow of light, maximising the energy they absorb. In the upper layers of the leaves the cells are tightly packed, rich in chlorophyll, abundantly connected with water-supplying vessels. In contrast, the lower layers of the leaf are spongy with air spaces (fed by the many pores in the lower epidermis), thus bringing carbon dioxide from the air to the photosynthesising cells. The leaves are exquisitely crafted, efficiently bringing together the things that are needed for photosynthesis to occur. Structure relates to function, as biologists often say. The 'bottom-layer' transcription factor I wrote of a few days ago contributes to the making of this functional structure by controlling the activity of genes.

Friday 12th March

HEAVY RAIN IN THE NIGHT

Today I witnessed what could have been the end of the plant. I had been looking at some raindrops on a blade of grass at the grave's edge. Seeing the pearly shine of them, thinking about how they looked hard in the light despite my knowing that they are soft, when I saw a movement at the periphery of vision. A slug. Crawling away from the foot of the marble wall, inching towards the thale-cress. Its body long and black, coated with mucus. At its head a pair of antennae waving, extending and retracting. A mid-body section rough, pitted like pumice. A tapering tail section, its surface marked by ridges parallel to the length. The slug was on course. Sliming along a path that converged on the plant. I waited, then watched the moment of contact. It began to feed, grazing first on one of the older leaves. Its mouth moving left to right and back again. The whole leaf shuddering, moving back and forth as the slug tugged at it with its rasp. And then the slug stopped feeding, began to crawl again, continuing in the same trajectory as before. Pointing straight as a dart at the centre of the rosette. This was dangerous. The thale-cress plant could afford to lose a bit of leaf. But damage to the meristem could be serious. Within a few seconds the slug reached the margin of one of the younger leaves and began to eat again. Destroying the very structures I'd been thinking about just minutes before. The architecture of the leaf, the different types of cell. I watched the destruction, as the slug fed upon the contents of the cells: the sugars made mere seconds previously, the proteins in the cytoplasm, the meat of the leaf. Gradually, the slug chewed down from tip to base, following

the shape of the leaf as it led to the centre of the rosette, getting ever closer to the meristem.

But suddenly, it ceased feeding and began to crawl in a new direction. I was surprised by this unexpected violation of a momentum that seconds previously had seemed as fixed as that of a ship at sea. It sailed away on a new course that quickly took it to the outer limits of this small universe, leaving the other plants untouched. Watching its departure, I experienced an involuntary shiver.

Why did I not kill the slug? In retrospect it would seem the obvious thing to have done. However, watching it as I did for the space of perhaps twenty minutes, I felt I couldn't intervene. That even though I wanted to, to have done so would have been against the spirit of this natural history. How could I accurately depict the progress of life if I interfered with that progress? If my diary is to reflect a real life, how could I act to tip the balance in one direction?

I made a quick check of the damage. An elderly leaf nibbled at: towards its end a patchwork of holes bounded by tough veins, an etched reminder of what was once filled space. A younger leaf eaten down to the place where it became a stalk, jagged flaps of the blade remaining. I left with a feeling of calm after crisis. And a sense of having witnessed an event of significance. The flow of energy from one form of life to another.

On, then, to the grey sky and brown landscape of March in Wheatfen broad. To the buff reeds, dry and angular. Quiet, stillness. The flight of birds. Cold. A longing for spring.

Saturday, 13th March

A REFLECTION ON DNA

Excitement at the coming of spring. First, morning gloom. Then, later, wonderful light and the occasional shifting blackness of shower-clouds. Much milder than of late. By the afternoon it felt as if spring had arrived within the space of a few hours. On the oak outside the window, the buds at the ends of the twigs swelling with movement one can almost believe to be visible.

Across the city by bus. Through the centre, the Saturday shopping throng. It suddenly seemed to me so strange that we all live our day-to-day lives largely unaware of why we are here, why this urban landscape and the trees and fields not far beyond it exist at all. We've always wondered at the nature of our existence, built myths around various explanations of it. And now that we understand at least a part of it all, we don't really know it.

What I mean, of course, is that we're here because of the molecular properties of DNA, the material from which genes are made. DNA, that linear molecule of two strands wrapped around each other – the structure known as the double helix. Each strand

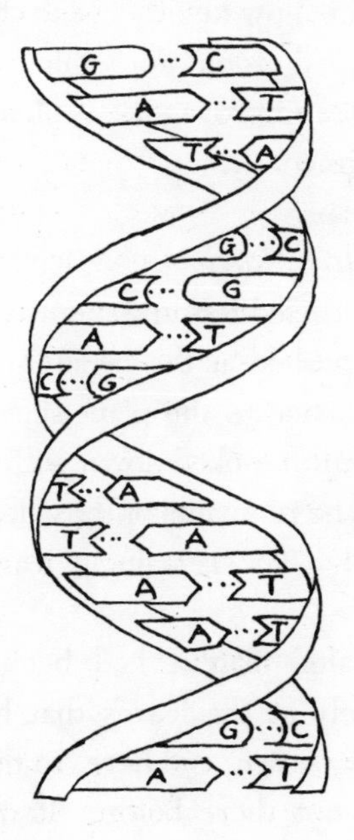

The DNA double helix.

consisting of a chain of four bases (written as A, G, C, or T); thousands and thousands of these bases in what seems at first sight to be random order: GATCGTGTTAACT and on and on. Trace it with your finger on the sketch. The two strands mirror images of one another, paired by the complementary shapes and fit of the bases from which they are made. Always G is paired with C and A with T, making the second example sequence strand CTAGCA-CAATTGA so that the sequence of the one fits snugly with the sequence of the other. Why don't we connect this knowledge with the reason we are here? Is it that we feel distance from it?

Later – to a concert in St George's, Colegate. A bare, austere interior. Cold. The Bach cantata that contains 'Jesu, joy of man's desiring', beautifully sung. A trumpet that shone, hovered sweetly above the movement below. I love the feeling of this piece. Of someone (Bach) who had a sense of belonging, the confidence that came from knowing how he fitted in to the rest of the world. It was of course a unified vision. Something I feel the lack of at present. Perhaps science improves the *accuracy* of our vision but at the same time fragments it?

Sunday 14th March

To St Mary's. I was up early and cycled fast in the new mildness. Sweat pricked at my chest as I pushed at the pedals.

There had been no further damage to the plant since last I'd seen it. Resting, catching my breath, I looked down with pleasure on the plant and its neighbours. The remnants of those leaves that had been eaten were still visible. The remaining leaves were untouched.

In fact, the plant is now safer than it had been before. Crouching down, I looked closely at the leaves that had been eaten. The severed edges are now brown, and have, to the touch, a horn-like toughness that was not there before. It marked a molecular change. New cell wall had been made, and new and old wall strengthened. Preventing evaporation, protecting the cells from infection by bacteria or fungal spores. Although it is possible that some infecting micro-organism may have entered the body of the plant before the wound was fully closed. Time will tell.

And there is a further line of defence. Being eaten has actually made the thale-cress's existence safer than it was before. The plant has been 'immunised' by its encounter with the slug. It sensed the chewing, felt it as it were, sent signals to the nucleus of the cells of the bitten leaf. Those signals caused an increase in the activity of genes encoding proteins of particular characteristics, and a resultant accumulation of those encoded proteins. These particular proteins inhibit the activity of enzymes. In fact, they inhibit the action of a specific class of enzymes that digest proteins. They inhibit the gut enzymes of slugs.

Was that sudden saving change of course by the slug caused by indigestion? Perhaps. Although it seems that it happened too soon after the slug had begun to feed for this to have been the case. Maybe the slug had simply developed a fancy for the flavour of some other kind of leaf. Nevertheless, the accumulation of inhibitors protects that leaf (at least what little remains of it) from further attack.

There is a yet more remarkable thing. It was not only the leaves that were actually chewed that were affected. The plant is now protected throughout its full extent. The first signals, those initially released by the slug's chewing, caused the production of further molecular signals. These second signals entered the vessels of the plant. Once there, they spread throughout the shoot, into the other leaves. Into leaves that were intact, that had not been damaged by the slug. These leaves now also began to accumulate inhibitors. When one leaf is eaten, all become protected. So, as I looked at the plant, I knew that the constitution of its whole body had been changed by the slug's attack. That it is now more defended in its entirety than it was before.

The slug's attack provoked enhanced defence. It is an example of the responsiveness of living things, of the power of evolutionary change to cause the adaptation of an organism in response to changes in its environment. In the last few days I've often found myself thinking of the thale-cress plant as if it were some wonderously complicated machine of many parts. That there is a stimulus and that the machine responds. That we will one day come to a full understanding of the biology of plants as being like the tickings and mechanism of a beautifully fashioned and complex clock. It is a pleasing image. And yet? And yet at the same time I feel a dimly understood disquiet that this image is too simple, too easy. That there are many aspects of the world, the weather for instance, that will never be explained in so predictable a way.

Enough. Today's was a snatched visit to this graveyard landscape. I had other things to do. I needed to get home. I turned to the bike, put a square of chocolate in my mouth, saw shining flints in the wall. Something came into my mind. About moving closer in to a place of safe haven. A port in a storm. A shelter. This natural history has got me involved. I care about it. As I left I saw that people were beginning to congregate in the church for morning service.

Monday 15th March

Spring on its way at last. Blown in over the last day or so by a mild, moist south-westerly. Everywhere buds are swelling. Green things expanding in the warmth. White blossom on the flowering cherry trees, some of it blowing off, flying and dancing in the wind. Sudden delicious wafts of blossom scent.

There's a profound sense of release. To walk outside in the mildness feeling the breeze in your shirtsleeves without needing the enclosing warmth of sweater or coat. To feel almost part of the patches of blue sky. And with all this comes the sudden realisation that I want to add a new thread to this journal. I still haven't seen a way forward for our laboratory research. I'm hoping that the addition of this new thread will help me to see where to go next. The plan is to write a recent history of the research done by my group.

We study growth, the growth of plants. I've always loved this phenomenon. Always wanted to understand it. Although the real goal, at least for me, is not simply better understanding. Nor the promise of future utility. I'm more motivated by the sense that understanding brings me closer to Nature. That there's a link between understanding and reverence. Although it seems a little naïve written this way. I hope I can find a better way of expressing it.

My research group studies the growth of plants by applying the logic of genetics. This logic works as follows: we investigate the growth of mutant plants, plants that don't grow properly. Mutant plants don't grow properly because they lack the normal function of a gene that enables them to do so. Studying the mutant permits deductions concerning the function of the normal gene.

Here's an important example. About fifteen years ago, I became particularly interested in a thale-cress strain that carries a mutant gene called *gai*. This strain is dwarfed. Compared with strains carrying the normal form of the gene, the mutant strain is dark green, has short stems and small leaves. It's dwarfed because of a slowed rate of cell proliferation (the cycle of successive enlargement and division of cells). It seemed likely then that the protein encoded by the normal form of the gene affects growth by controlling the rate of cell proliferation.

Another thing that affects the growth of plants is a hormone known as gibberellin (a name derived from that of a fungus called

A comparison of normal and *gai* mutant thale-cress plants.

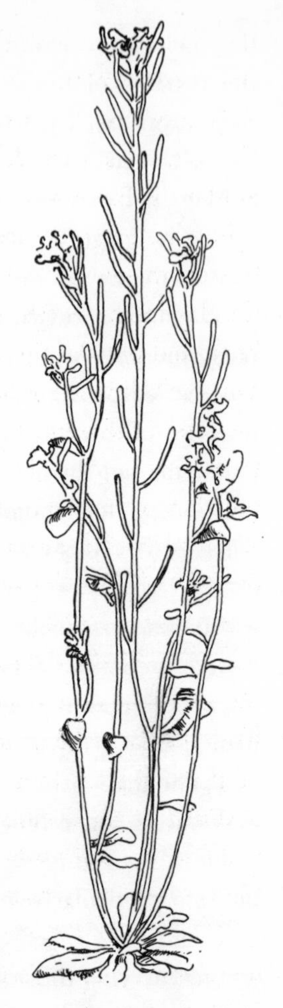

Gibberella that produces gibberellin in large quantities). Normal plants grow relatively tall because they make enough of the gibberellin growth hormone to grow normally. But some kinds of mutant plants are dwarfed because they lack the ability to make enough hormone. These gibberellin-deficient mutant plants look very like *gai* mutant plants. But *gai* mutant plants are not gibberellin-deficient. On the contrary, *gai* mutant plants make gibberellin, and are dwarfed because their cells have a reduced ability to respond to it.

So we knew that the *gai* mutation affected growth. That it did so by altering the responses of plant cells to gibberellin. What we didn't know was the nature of the normal form of the gene, of the protein that that normal gene encodes, or how, when the gene is mutated to the *gai* form, this changes the responses of the plant's cells to the hormone. To find these things out, we needed to isolate the gene that was mutated in *gai* mutant plants.

Tuesday 16th March

ON FLOWERING

This morning the mildness continues. The wind pushing the clouds at a fast pace. It was hard work cycling into it on the way to St Mary's. But when I got there the celandine flowers were bright yellow in the graveyard. Birds trilling in the chestnuts. The spring flowers making a mark in the mind. And I wondered this morning if today might be the day when the thale-cress begins to flower? I knelt and put the tip of my finger into the crown of the rosette. You can tell when a thale-cress plant is about to flower by feeling the top of the stem. Long before you can see flower buds you can feel them as a roughness. The pointed tops of the buds abrading the skin of fingertip. But what I felt today was smoothness.

Of course the actual decision to flower is taken long before the flower buds are apparent. Is it possible, then, that that decision was taken today? That today a final switch was tripped? A switch in a chain of switches that all need to be thrown before the plant can enter the next stage of its life-cycle. With outer leaves ragged from the past winter, still holed by the hunger of the slug, could it be that the plant today became a flowering plant? If so, it is now destined to make flowers and leaves and roots, whilst before it made leaves and roots only. Flowers that will soon shoot up into the air on a stem that will grow out of the centre of the rosette.

The transition to flowering begins with a transformation of the meristem cells, the ball of cells at the growing tip of the shoot. With that transformation, the meristem stops being a vegetative meristem and becomes an inflorescence meristem. And just as the vegetative meristem makes leaves in a spiral of bulges around itself, so the inflorescence meristem makes a spiral of flowers. This change in identity is caused by factors inside the plant and by others from the world outside it. Inside the cells of the plant are genes that, when activated, induce inflorescence-meristem identity. These genes encode transcription factors, proteins that activate further genes. When these further genes are activated, the proteins they encode in turn work together to transform the vegetative meristem into an inflorescence meristem. This is known from studies of the effects of mutations that stop the inflorescence meristem-identity genes from working. For instance, a mutant plant carrying a non-functional form of one such gene, called *LEAFY*, continues to make leaves at a time when it would, if all were normal, be making flowers. The transition fails in the mutant: the vegetative meristem does not become an inflorescence meristem because that transformation requires the action of the further genes normally activated by *LEAFY*.

LEAFY is an internal thing. But the ultimate activator of *LEAFY* is the world outside the plant. Indeed, the more I think about it, the more I start to wonder if what is inside and what is outside the plant are really distinct things. About whether the identification of the plant as a separate entity is a construction of mental convenience that enables us to see what we see within the context of a pattern of meaning but that is an artificial construction none the less.

Wednesday 17th March

HOW THE OUTSIDE WORLD CONTROLS *LEAFY*

The activity of *LEAFY* is dependent on the variations in light and temperature that are the consequence of seasonal progression. Last autumn, as the thale-cress plant was constructing its rosette of leaves and growing into winter, an additional gene, named *Flowering Locus C* (or *FLC*) was activated. *FLC* encodes a protein, FLC. FLC is a repressor of flowering and prevents the vegetative meristem from changing to the inflorescence state by blocking the activation of *LEAFY*. In the presence of FLC, the *LEAFY* mRNA isn't made, thus preventing the formation of the LEAFY transcription-factor protein, and thus halting flowering. The activation of *FLC* prevented the thale-cress plant from flowering inappropriately during the winter.

FLC is itself activated by yet another protein, a protein encoded by a gene called *FRIGIDA*. Somehow, the level of *FRIGIDA* mRNA is controlled by time and by temperature. The plant measures the length of time during which it is exposed to the cold. The longer the exposure, the lower the level of *FRIGIDA* mRNA becomes. Thus, by the end of last winter the plant's *FRIGIDA* mRNA level had become very low. It is an exquisite property of this system that a plant can remember that it has been through the cold, even when the weather becomes warmer. This memory prevents *FRIGIDA* mRNA levels from rising again. As they fall, so FLC activity also falls, *LEAFY* is no longer repressed, and flowering is promoted. So now I am wondering if, on this mid-March day, and perhaps for several weeks past, the levels of *FRIGIDA* mRNA have been so low as to make the FLC levels in the thale-cress plant insufficient to prevent flowering.

prolonged cold ⊣ FRI → FLC ⊣ LEAFY → flowering

How a period of cold promotes flowering. Prolonged cold inhibits *FRIGIDA* mRNA, or *FRI*. Since *FRI* promotes *FLC*, the prolonged cold inhibits *FLC*. Since *FLC* inhibits *LEAFY*, the cold-inhibition of *FLC* promotes *LEAFY*. Increased *LEAFY* promotes flowering.

However, there is more to be done before the plant will actually flower. Removing FLC is necessary for the activation of *LEAFY* and the initiation of flowering. But it isn't enough. There's another step.

Thursday 18th March

HOW DAY LENGTH TRIGGERS FLOWERING

For the next step towards flowering, the plant needs a positive stimulus. Something that specifically activates *LEAFY* rather than simply removing a block to its activation. Could it be that today that switch was finally tripped, flipped by the length of the day?

Actually this next switch works much like the *FRIGIDA*–FLC switch. By the convergence of internal and external things. And again the internal thing is a transcription factor, a protein that acts as a regulator of genes. This protein is encoded by a gene named *CONSTANS*, so called because mutations in it cause plants to flower at more or less the same age independently of the length of the day. The thale-cress plant on the grave, a normal plant, doesn't behave in this way. It uses *CONSTANS* to enable it to flower in the long days of spring and summer rather than during the short days of winter.

The expression of *CONSTANS* is controlled by an internal clock. A mechanism that enables a plant to measure time, that marks a period of twenty-four hours and is re-set by the dawn. The levels of *CONSTANS* mRNA are low at dawn and gradually increase throughout the day. This spring, as the days have lengthened, so *CONSTANS*-mRNA levels have become progressively higher and higher at successive dusks. During the night the levels remain stable, but they fall to starting levels at dawn. This fluctuation is reflected in the level of the CONSTANS protein encoded by the mRNA. But the dusk level is not in itself sufficient to trigger flowering. CONSTANS needs to be activated by light. A plant detects light through the action of light-sensitive proteins that work as photoreceptors. Absorption of light by these photoreceptors sets in motion a chain of events that leads to the activation of CONSTANS. Once there is sufficient activated CONSTANS, the *LEAFY* gene is activated, and the meristem is transformed from a vegetative to an inflorescence meristem. The plant's flowering is induced by the convergence of two independent things: the day being long enough to make sufficient CONSTANS, and the presence of the light needed to activate it.

This is how that final switch will be thrown. The thale-cress plant will then cross a threshold. Everything will change. Yes, the meristem in the shoot tip will continue to make bulges with the outgrowth of cells on its flanks at regular intervals. Yes, those bulges will still be arranged in a spiral. But, after the transition, they will be different bulges, bulges destined to become floral meristems rather than leaves. A refrain is exemplified here. That nature constructs by modification of what was previously there. Rarely is anything completely new. Flowers and shoots are related by derivation.

And it is of course important that the plant flowers at the right time with respect to the progression of the seasons. The essence of the last three diary entries is that every plant detects and responds to changes in the world of which it is a part. It does so via a hierarchy of gene control, with genes controlling other genes. 'Genes controlling other genes' is becoming something of a litany here.

Friday 19th March

FIRST EXPERIMENTS WITH *gai*

Beautiful spring sunshine this morning. A woodpecker heard trilling and tapping at a trunk in the woods on the way to work.

To return to the *gai* gene and how it was cloned. It took us many years to clone that gene. Long years of hard work. But it was well worth it. The name of the mutant form is written *gai*. The normal form is written *GAI*. The mutant form confers dwarfism. Yet another convention operates here. The normal form (*GAI*) is written in capitals, the mutant form (*gai*) in lower-case letters. And, as described previously, the gene is named after the mutant form, in a way that describes what happens when it fails to work as it should.

One of the properties of *gai* is that it is genetically dominant (more about genetic dominance in a minute). Thale-cress is a 'diploid' organism. By which I mean that each of its cells contains two genomes, one of maternal, the other of paternal, origin. So the cells of thale-cress contain two copies of each gene, one copy from each parent. A plant that has two copies of the *GAI* form grows tall. A plant that has two copies of the *gai* form is dwarfed. A plant that has one copy of *GAI* and one copy of *gai* is also dwarfed, although not as dwarfed as a plant that carries two copies of *gai*. And the fact that a plant carrying one copy of *gai* is less dwarfed than a plant carrying two was key to our first real experiment, as I will describe shortly.

The commonest kinds of gene mutation are those which prevent genes from working. For the gene to be damaged in such a way that it no longer functions. Such mutations are usually found to have genetic properties that are said to be 'recessive' as opposed to the 'dominant' properties outlined above. The meaning of 'dominant' and 'recessive' can be thought of in the following way. Usually, a plant that contains one copy of the normal form of a gene and one copy of a non-functional mutant form of that gene looks normal. The effect of the mutant gene is said to be recessive. In contrast, a dominant mutant gene is one that manifests its effects irrespective of the presence of a normal form of the gene. Such a dominant mutation is less likely to be due to simple damage, to the mutant form being one that no longer functions. This dominance is exactly what is observed with *gai*. So we guessed that the *gai* form is not simply a non-functional gene. We imagined that perhaps *gai* makes a product, a protein, but a protein that is somehow subtly altered in the way it works.

In the way it controls the growth of a plant. According to this hypothesis, *gai* is a gene of altered function but not a destroyed gene.

One way of testing this hypothesis was to look for mutants in which the gene had been changed again. We made a prediction: that if we took the altered-function *gai* mutant and mutated it again, we'd be able to totally destroy it. As a result, this destroyed gene, rather than making a changed protein, would make no protein at all. This destruction might then be revealed because of its effects on growth.

How do you change the structure of a gene when you don't even know what that gene is? Can only infer its existence from its effects on the growth of a plant? Actually, geneticists have been doing this for years. By making use of mutagenic agents, agents that alter the structure of DNA. Treatment with these agents affects, more or less at random, a few of the genes in the genome. To find mutants, you look amongst a population of treated organisms (or their progeny) for the rare individuals that look different from the rest. Different in a way that might be attributable to an alteration in gene structure.

Looking back from a time when genes are concretely understood in terms of their being segments of DNA, it's amazing how the early geneticists were able to reveal so much about the structure and function of genes from mutation experiments performed in this way. In fact the depth of knowledge those pre-DNA geneticists were able to achieve was the result of remarkable powers of observation and imagination. Such penetrating insights into the ordering of genes on chromosomes (which we now know to be strands of DNA), into the fact that genes encode enzymes (proteins), into the activity of genes, and into genes that are mobile and move from one site in the chromosome to another.

So here was our first experiment. It was rooted in those classical experiments, culturally a part of the world of pre-DNA genetics, and based on a prediction: a 'what-if' imaginative step into the unknown. The prediction was that *gai* is an altered gene that makes an altered protein. That the altered protein retards the growth of plants. If we were to destroy *gai*, then we might guess that plants carrying the newly destroyed form of the gene might grow tall rather than dwarf.

To begin the experiment we took 60,000 seeds, each containing an embryo made of cells containing two copies of the mutant *gai* gene, and exposed them to gamma rays. Gamma rays are powerfully mutagenic. They are highly energetic and damage DNA when they collide with it. Each of the genes of each of the cells of each of those 60,000 embryos was a possible target for this damage. And I remember thinking how it seemed almost strange that the seeds looked the same after the treatment as they did before it. Despite the fact that, inside, at the level of the DNA molecules from which their genes were constructed, they were fundamentally changed.

We planted the 60,000 seeds and waited for them to germinate. But germination was slow. The radiation dose is hard to get right, and the sensitivity of seeds to radiation varies with respect to their moisture content, a thing that is difficult to estimate. Too large a radiation dose can kill the seeds outright. Had we got the dose right? Perhaps the seeds were all dead.

But eventually they germinated. And when they did, the resultant seedlings showed symptoms typical of radiation damage. Their first leaves were slow to appear, and those that did were often distorted in shape. Many seedlings died at this stage. But the majority continued to grow.

What was happening during that faltering period between initial germination and seedling establishment? As I wrote on 29th

February, the cells of leaves are derived from the cells of the shoot-tip meristem. But the meristematic cells of these seedlings had been hit by gamma rays, their DNA damaged as a result. Some of these cells would have suffered damage great enough to have killed them. But other cells would have been less affected. Cells have the capacity to repair damaged DNA, although sometimes the correction is imperfect and perpetuates alteration in DNA sequence.

Classical experiments, in which sections were cut out of meristems, had shown that the meristem has fantastic powers of regeneration. Even small portions, consisting of just a few cells, are capable of regenerating entire new meristems out of themselves. So perhaps, in our experiment, the delay in seedling emergence corresponded to a period of restructuring. Of viable and semi-viable cells rebuilding the meristem out of themselves and their descendants, replacing those cells that had been killed outright by the radiation.

Gradually, the damaged meristems repopulated themselves with viable cells. These viable cells were to be the source of the plant shoots to come. We knew a fascinating thing about the plant bodies these cells were soon to build. That a substantial proportion – whole stems, branches, leaves, and flowers – of the entire body of a plant can ultimately be derived from a single cell in the meristem. That the shoot can be considered as a jigsaw puzzle consisting of a few sectors of tissue, the cells comprising each sector all being the descendants (via cell proliferation) of an original single cell of the meristem. This fact was key to the success of our experiment.

In every seed there is an embryonic plant. In our experiment, the 60,000 seeds contained 60,000 embryos. Because the target of 30,000 genes (the approximate total number of genes in the thale-cress genome) is very large, the chance of any one individual gene (in our case *gai*) being damaged by the radiation was relatively small. But imagine that one of the *gai* genes in one of the one or two hundred meristematic cells of one of the 60,000 embryos was damaged by the radiation – let's call that gene *gai-d* for 'damaged' – causing this particular cell to carry one copy of *gai* and one of *gai-d*. Imagine now that this cell was destined to be the founding cell of a large sector that formed a substantial part of the shoot that was growing. We could make a prediction about what the stems and leaves that grew within that sector would look like. We already knew that the growth and proliferation of cells containing two functional copies of *gai* was slower than that of cells containing just one. So we could guess that stems containing *gai* and *gai-d* might grow taller than all of the rest of the stems (containing two copies of *gai*) in our experiment.

This is exactly what we were looking for. Taller stems. Every day we went to the greenhouse to see how the plants were growing. With a blended sense of excitement and impatience at the pace of growth. And then we found some. Thirteen stems that seemed to grow taller than all the others. Thirteen from sixty thousand! It took a lot of looking. A lot of initial uncertainty. Like watching the development of a photographic print. At first we thought we could see something. Would discuss it amongst ourselves. Perhaps this one was a little taller than the others. Perhaps not. The following day we mightn't be so sure, couldn't see the difference so clearly. But as the days passed and the plants grew, our certainty strengthened. We discarded a few we'd been wrong about and settled on the thirteen stems.

So far, so good. But we had no real confirmation that the taller stems were in any way related to a mutation in *gai*. What we'd observed was a transient phenomenon in a few plants. Would the change in growth be passed on in subsequent generations? Could we convince ourselves that what we'd seen was to do with *gai* and not to do with one of the plants' other 30,000 genes? These questions required further experiments.

The thirteen stems carried flowers, and we let those flowers self-pollinate. The question was: would the *gai-d* genes that we guessed existed in the cells of the long stems be transmitted through sperm and egg to the next generation of plants? Basic genetics predicted that self-pollination of flowers borne on taller stems containing both *gai* and *gai-d* would yield seeds containing embryos containing *gai/gai*, *gai/gai-d*, or *gai-d/gai-d* in a 1:2:1 ratio (the ratio first described by Gregor Mendel in his classic experiments with peas). Would we see a distribution of different plant types amongst the progeny of the taller branches that corresponded to this prediction?

The result was exhilarating. The progeny families of four of the thirteen tall stems behaved just as we had predicted. The memory of my initial sighting of this is as clear as if I'd seen it this morning. That first tested family consisted of twenty-one plants: five that were dwarfed and dark green (as expected for plants made of cells that were carrying two copies of *gai*), eleven that looked as expected for plants carrying only one copy of *gai*, and, most excitingly, five that grew as tall as normal plants (although of course they weren't normal but carried two copies of *gai-d*). 5:11:5 was roughly 1:2:1.

Writing about this now, more than ten years after the event, I still feel the excitement it generated. There is a certain thrill in making a prediction about the world, testing that prediction, and finding that the result is in accord with the prediction. Better still, we'd found out something new about how plants grow. Could construct a genetic scenario that fitted our findings. It suggested, first, that normal plants contain a gene (*GAI*) that controls growth because it encodes a growth-regulatory protein (GAI) that responds to gibberellin. Second, that a mutant form of that gene, *gai*, encodes a mutant protein (gai), whose altered properties render it insensitive to gibberellin. Third, that *gai*, as we had now shown, can be mutated again. Thus making *gai-d*, a new form of the gene, a form that has no function rather than altered function. The *gai-d* form causes plants to grow tall and to look like normal plants (like plants containing the original *GAI* form).

This was a first step towards our current understanding of the DELLAs, a family of proteins which we now know to be utterly fundamental to the control of the growth of plants. They have been the major focus of our work ever since. But here I'm getting ahead of myself. The important thing for now is to emphasise that this first step also provided a key, a way in, to the problem of how to isolate the *GAI* gene itself. Of how to separate it, clone it away from the 30,000 other genes in the thale-cress genome. To find it, a needle in a haystack of DNA.

What started with this first step has ultimately helped us to understand the growth of all plants, not just that of thale-cress. The oak outside my study window, the chestnuts in St Mary's churchyard, the reeds in the fen. But why do I feel a tinge of awkwardness at these last sentences? They are almost certainly true. There is something about scientific culture that demands precision, that restrains the mind in its leaps to connect one thing to another to another. It is known that GAI-like proteins are crucial regulators of the growth of many different kinds of plant. But it is not formally proven that these proteins regulate the growth of reeds or oaks. Nevertheless it is vastly more likely that they do than that they don't.

Saturday 20th March

THE STRUCTURE OF ROOTS

Jack woke me just before dawn. Said he was feeling sick. Thankfully it passed, and in a while he began to feel better. I put him back to bed, and he was soon asleep again. But now *I* couldn't sleep, so watched the dawn. Slowly the light returned, pale beginnings seen through smarting eyes. The upper sky streaked with pink. And then my attention was grasped. By a sudden fluting twitter from a sparrow in the hedge, and by the chasm of silence that followed. Twitter. Silence. A declaration. Then sounded again. This time followed by another, fainter, further off. Perhaps from the oak. Then I heard the first fanfare of a blackbird in the bushes, soon followed by an answer from elsewhere in the garden. And then another again, and all the while a background to the more defined sounds, a scratching restlessness of squeaks and warblings, arose from the silence. As their strength grew, the various voices became intertwined until, within minutes, I was immersed in a cacophony through which the verses and responses of the blackbirds cut sharp as the sounds of trumpets.

During that jagged symphony I realised that I'd forgotten something from the thale-cress plant's story. That in all my excitement with the switching, the changing of the shooting part to a floral from a vegetative state, I'd neglected to tell the tale of the rooting part.

Perhaps this omission is not a surprise. After all, the roots are on the dark side of the boundary between air and earth. We all of us see the trunks and branches of the tree but not the roots beneath.

I dressed and went again to the edge of the graveyard landscape. Envisaging now what was happening underneath it rather than what could be seen above. Imagining the roots. They've been growing throughout the life of the plant. Gently penetrating the earth, navigating between the particles of rock within the soil. By now their tips are likely to have reached a depth of a foot or more, but unlikely to have reached the bones buried further down.

Let's write of roots in the geometry of their dimensions. They are cylindrical organs. First the dimension of length, the long axis. At the tip of the root there is a meristem, the source of all the cells from which the root is made. Further back from the meristem is the region where the cells elongate and where they begin to assume their various cellular identities.

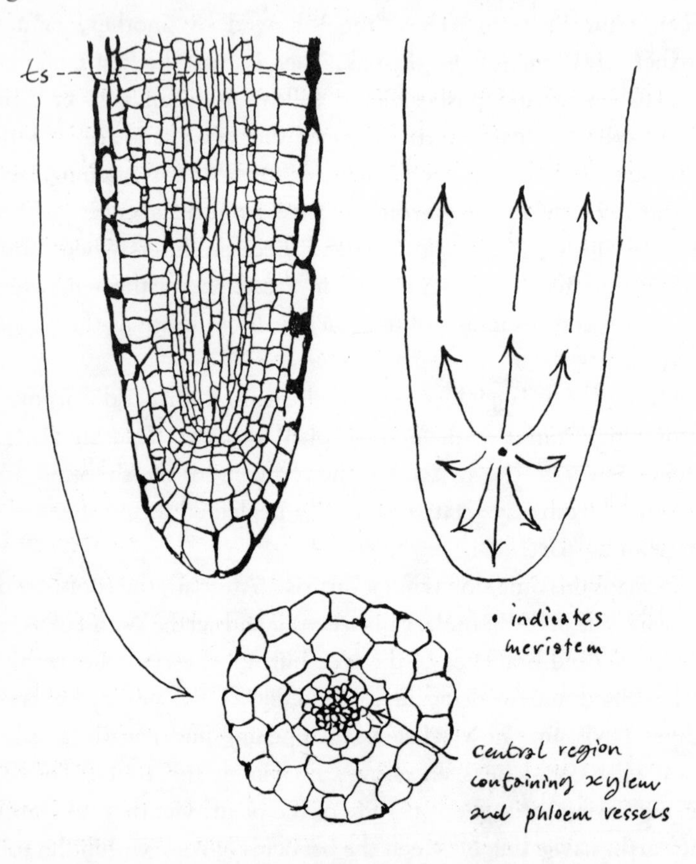

A thale-cress seedling root in longitudinal section (*above, left*), showing the paths of flow of cells leaving the root meristem (*above, right*), and in transverse section (*ts, below*).

There are, of course, similarities between the growth of roots and the growth of shoots. Roots, like shoots, grow by the creation of new cells in a meristem close to their tips. As the cells are pushed out of the meristem they form the shaft of the root. The force of cell expansion and division in that shaft pushes the tip of the root forward through the soil. And as the cells divide and elongate they become specialised, assuming one of several different identities.

The second, cross-sectional dimension of the root reveals it to be radially symmetrical. It is a cylinder made of concentric layers of cells. In the very centre are the xylem vessel cells, which enable the water absorbed by the root to flow upwards to the shoot, and the phloem cells, which enable the sugars and other nutrients to flow downwards from the shoot to the root.

Sunday 21st March

THE CLONING OF *GAI*

The fourth Sunday of Lent. There was heavy wind and rain in the night, and still more this morning. How do we see rain? As a falling blanket, as a collection of points, as individual globes with a surface and a centre (little worlds of their own), as an aggregation of molecules of H_2O? How do we fuse these visions into one?

Today it's warm. A strong westerly wind keeps the trees in constant motion, the rise and fall of roar and moan. Humid mild air from the Atlantic. A sense of enormous energy. The trees squashed from one shape into another. Excitement.

In the garden, many things now in flower or in bud: primroses, pulmonaria, daffodils, narcissi, hyacinths, the first forget-me-nots. All a-jiggle in the gusts. The box hedge is beginning to make those pale green new leaves I so love, the ones that stand out against the darker green of last year's growth.

The hazel buds now expanding. At the tips of twigs and at

intervals along them. Wrapped in regular overlapping scales of bud-cases. I'll study them over the next few days.

To continue with the story of the cloning of *gai*. We'd shown that it's possible to destroy the *gai* gene, and that its destruction was detectable. Not because we could see the gene and directly observe its destruction, but because we could see the consequences of its destruction. A plant carrying *gai* is dwarfed, whilst a plant carrying *gai-d*, the destroyed form, is tall. Detection by inference.

It was that simple observation that opened up the route to the isolation of *GAI*, the identification of the piece of thale-cress DNA that was the gene itself. We wanted to isolate *GAI* because we thought that its isolation would in turn provide a much clearer vision of how the protein (GAI) it encodes controls the growth of plants. And of how the mutant gai protein retards the proliferation of cells and makes plants dwarfed.

But the problem was formidable. All of the 30,000 genes (or thereabouts) in the DNA of thale-cress are very similar to one another, just bits of a huge strand of DNA. Nothing to clearly distinguish one from another. So how were we to find the one in 30,000 that was *GAI*?

We were aware of a potential way forward. A route that others had previously used to isolate genes. But we knew it to be an unreliable method, a method that doesn't always work, and whose effectiveness varies according to the particular gene that is the focus of attention. Taking this approach was risky.

Its scientific name is 'transposon-tagging'. It uses a small piece of DNA known as a transposon that has a rather startling property. Transposons jump from one place within a larger DNA molecule (a whole chromosome, say) to another. They are capable of inserting themselves into other DNA molecules, of interrupting the original sequence of the DNA at the site into which they've inserted themselves.

This insertional property creates gene mutations. If a transposon inserts itself into a gene, it can destroy that gene's function. It was this mutagenic capacity that was key to using a transposon to isolate *GAI*. We'd already shown that disruption of *gai* with gamma rays results in plants that are tall rather than dwarfed. Could we replicate this finding, now mutating the plants via the insertion of a transposon rather than with irradiation-induced damage? If we could, then that insertion would allow us to isolate the gene, because the DNA of the transposon itself was already isolated. Insertion of the transposon into *gai* would 'tag' it, making the isolation of *GAI* possible.

But would it work? It might, we thought. But then again, it might not. There were many potential problems along the way. The idea was exciting, but almost certainly there would be frustrations. And perhaps this might in the end turn out to be a blocked path, a path we walked down for several years unable to see the wall at the end that would finally bar the achievement of our desired destination.

So we set out on this adventure. With apprehensive excitement. Transposons tend to insert themselves into sites that are relatively close to the site from which they jump. So we began with a plant carrying the transposon in a place we already knew was only a short distance from *gai*. Jumping is driven by an enzyme called the transposase, an enzyme that catalyses the excision (jumping out) and insertion (jumping in) of the transposon from its old and to its new sites. The whole process is like asking an archer who is blindfolded and has only a vague notion of the location of the target to shoot a bullseye. Although we'd enhanced our chances of hitting the bullseye by placing archer and target a small distance apart, still the archer could be expected to miss the bullseye most of the time.

Once again, we screened many thousands of *gai/gai* plants, looking for tall stems. Stems that carried one copy of *gai*, and, we hoped, a new version that we'd call *gai-t* (for 'transposon insertion'). When flowers from these longer stems were allowed to self-pollinate, we imagined we'd find that one quarter of the progeny would be tall, normal-looking plants (rather than dwarfed). That these tall plants would contain two copies of *gai-t*.

To begin with, our confidence was strengthened by what we saw. We found tall stems. Each representing an independent mutagenic event, each presenting a unique chance of being what we were after. Even better, when we planted the progeny of these stems, four separate families contained roughly one quarter of plants that were tall, plants that potentially carried a *gai-t* gene within their cells. We optimistically labelled these potential mutant genes from *gai-t1* to *gai-t4*.

But then the problems began. Our next step was to determine if the DNA of the new putative *gai-t* plants carried a transposon in a new position, as would be expected if a transposon had inserted into *gai*. But when we did this, we could see nothing new. Not one of the putative *gai-t* plants carried a transposon in a new position.

In truth we had all along been aware that this might happen. There was a ready explanation for it. At the time that we were doing these experiments, thale-cress transposon-tagging was an infant technology. For instance, the level of transposase activity optimal for getting transposon insertions into target genes wasn't known. But it was known that too high a level of transposase could be a problem. If the transposase is too active, it can cause a transposon to jump out of a gene just as soon as it has jumped in. Each time this happens there's a risk of residual damage to the gene. This, we guessed, was what had happened. The transposon had hit *gai*, damaged it, and then gone again. Because the *gai* gene was damaged, it caused the plant containing it to be tall rather than dwarfed. But because the transposon was gone, the gene was not tagged, and we had no way of isolating it.

We were frustrated. If only we'd used a less powerful transposase source! And of course we still could. But our lack of certainty that the power of the transposase source really was the root of the problem inhibited us from taking this path for a while. It meant going back almost to the beginning. Months of work wasted. Months more to come with no certainty of success.

But of course we did try again. Using a less powerful source of transposase. There was nothing else to be done. This time we only isolated two tall stems, from many thousands more plants than we had screened previously. It makes sense, we thought. Less transposase, less transposition, fewer events. Both mutations came through to the next generation, making tall plants in a quarter of the progeny. We called these mutations *gai-t5* and *gai-t6*.

We checked the DNA of *gai-t5* first, hardly daring to hope. The result: renewed frustration. Again no detectable transposon. Because of the growing darkness in our minds, we very nearly abandoned *gai-t6*. We thought it would just be the same depressing story all over again. But we were wrong. With a final effort we searched for a transposed transposon in *gai-t6*. And we found one! With the first sight of that result came the feeling that we'd crossed a threshold into a whole new world of possibility. It was a moment of breakthrough. *gai-t6*, a plant in which *gai* had been inactivated, contained a transposon inserted into a new site.

Frustration swung in seconds to elation. We had a strong chance now, and we knew it. But first we had to persuade ourselves that we really had tagged the *gai* gene. All that we knew so far was that we'd obtained a plant in which *gai* was potentially inactivated (a plant that was tall rather than dwarfed) and that this plant contained a transposed and re-inserted transposon. That transposon could be inserted anywhere throughout the entire genome of the thale-cress. We had no proof that the inactivation of *gai* had been caused by the re-insertion of the transposon, nor any indication that the transposon was now inserted into *gai* itself.

So we used a technique that enabled the purification of the precise region of DNA into which the transposon had newly inserted, and obtained the sequence of this DNA. By scanning through that sequence, we were able to find a stretch that had the expected characteristics of DNA that encodes protein (a segment of a so-called 'open reading frame', something I'll explain later). Furthermore, this open reading frame was interrupted in *gai-t6* plants by the insertion of the transposon. This was very encouraging. We were excited, hopeful, and happy. These new results showed that the re-inserted transposon in the *gai-t6* line had interrupted a gene. And because that specific interruption had occurred in the same line in which *gai* had been inactivated, it suggested that this newly discovered fragment of open reading frame was part of the protein-coding region of *gai* itself. Yet still the case was not proven.

Proof came with our subsequent experiments. Now that we knew the DNA sequence of the likely *gai* open reading frame we could use it to determine that of the open reading frame from the *gai-d* lines. The lines made previously, that contained a *gai* gene inactivated by gamma rays. Gamma rays cause characteristic damage to DNA. If this new candidate open reading frame was indeed *gai*, we would have expected it to have been damaged in these *gai-d* lines. The open reading frame of the first *gai-d* line we checked was clearly damaged. And so were all the others. It was proof, and we were delighted. At last we had done it. And from here we could progress. We'd scaled a height from which we'd be able to see further.

This proof came in the early summer of 1996. Seven years previously I had returned to England from California with the objective of cloning *GAI*. Now, at last, that objective had been achieved.

Monday, 22nd March

HOW ROOTS NAVIGATE THEIR PROGRESS

Spring advancing. The weather prickly with warm sunshine and sharp showers. A torrential downpour in the night. This morning the pungent smell of sodden soil excited thought. Horse-chestnut buds plump, shining with vulgar stickiness.

The hazel buds are opening. Less compact than they were. Bud scales lifting like the wing-cases of a beetle before flight. Revealing the tiny, ridged velvet-green leaves inside.

There are mats of yellow celandine everywhere: in the garden, the fen, the churchyard. The petals as if varnished, a shining ring around a doughnut of anthers around a pincushion of fused carpels. I picked one and pushed my thumbnail into the stem, separating its material into halves, watching the sap ooze from the cut. And as I did so I imagined how it was at microscopic level: my nail penetrating the stem, pushing all the way through from one side to the other, separating the columns of cells, dividing them into halves. The cells touched by the edge of my nail most likely squashed or broken, their contents released, leaking out. I began thinking again about the difference between what is alive and what is not, how the one seems to flow into the other such that there's no clear boundary between them. That it's hard to see where celandine begins and earth ends.

A day or so ago I wrote about the roots of the thale-cress. Yet this morning, when I reread what I'd written, it seemed to be lacking in dimension. As though the plant was simply growing on its own, the roots building themselves without reference to the rest of the world.

So I decided to extend that limited picture. Once again I stood at the edge of the grave, neck bent as I looked down on to the plant. Once again the thale-cress roots were extending beneath my feet. This time I reflected on how each root tip is sensitive to the world. Sensitive at once to the broad dimensions of the globe, and to the fine-grain, crumb-by-crumb texture of the soil itself.

First, to take the broad dimensional. The navigational capacity of the tip of the root. Roots sense where they are growing, and how the direction in which their tips are travelling relates to the gravitational vector. To the line that joins the centre of the earth to the point we occupy on its surface. Last autumn, the first seedling root pushed into the soil and grew straight downwards towards the earth's core. It did so because there are cells in the very tip of the root that tell it which direction to grow in. These cells contain grains of starch within their cytoplasm. Because they're more dense than the rest of the cytoplasm, the starch grains are pulled by gravity, and collect on the bottom of the cell. Somehow, the cell can sense where those starch grains are. Can use the positions of the grains to determine its orientation relative to the gravitational vector. The cells that contain the starch grains communicate this information to cells further back in the root shaft, the cells that actually do the growing. If the root tip is displaced, the starch grains shift their position, and the cells containing them sense the change. As a result, the growth of the cells on one side of the region of the root that normally drives growth is restrained, whilst the growth of cells on the other side is not. Accordingly, the root curves, until its growth becomes once again parallel with the gravitational vector.

So roots are controlled by the world within which they grow. Their growth trained to the gravitational vector through their ability to sense the direction of the gravitational force. They have a sense of orientation, of place and position. They know where they are.

But I digress. What of the second subject, the sensitivity of the root to the fine-grained features of the world? Imagine the growth of one of the thale-cress's roots, perhaps a foot below the ground. The tip is being pushed forward by the growth of the cells just behind it, finding its way between the particles of stone, decaying fibres of dead vegetation, bits of stick, and grains of sand that make up the soil. Last night's rain is swelling the cells of the root, making the growth fast and vigorous. But then the tip collides with an obstacle. A large stone. There is no way through. The cells of the elongation zone behind the tip are still pushing it forward, squashing it up against the stone's hardness. The root deals with this problem by responding to it. The cells of the tip begin to make a signal, a simple hormone called ethylene. A molecule constructed of two atoms of carbon and two of hydrogen (written as C_2H_2), its production triggered by the stress and pressure the cells are experiencing. Further back in the root, the cells of the elongation zone contain a protein receptor that detects the presence of ethylene. When the ethylene generated in the tip reaches the cells of the elongation region, it binds to the receptor, changing its shape, triggering a chain of signals inside the cell. This signal chain reduces the rate of expansion of the cells of the elongation zone, thus slowing the growth of the root and reducing the risk that the collision will damage the delicate cells of the root meristem. As the root grows more slowly, so it begins to feel its way around the edge of the stone. So a growing root is a sensitive thing, picks its way through the texture of the soil, finds the best path.

As I stood there, looking down at the ground, there arose in my mind the unbidden thought that this plant to which I've recently paid so much attention will soon be dead. That this object of coherent complexity, this sensitive root of countless components, is a transient thing.

Later, at work this afternoon, I felt rather flat. Unexcited and uninvolved. Perhaps I'm coming down after all the excitement with the paper that was recently accepted. Everything seems to be moving slowly. Like the growth of the thale-cress plant. Slow, steady, uneventful. And it's odd how one can see the same thing in such different ways. Sometimes, as this morning, the growth and the spring's progress seem so precipitous. Sometimes, as this afternoon, they drag. Sometimes so robust. Sometimes so fragile.

And still I can't see where to turn with our research. I'm getting bored with writing this over and over again, of finding different words to say it. 'Still stuck' sums it up. But there's a sense of mounting pressure. Sooner or later something will give, a new insight will break out.

Wednesday 24th March

THE EXPANSION OF LEAVES

Yesterday to an auction at Banningham Old Rectory, in lovely spring sunshine, occasional showers crescendoing from patter to roar on the fabric of the marquee. But uncomfortable heat inside when the sun came out. And I wasn't very successful. Particularly taken by some of the 'antiquities': a medieval stone lion-face and the most beautiful Roman marble *cinerarium* casket, the stone such a creamy colour and the carving austere, solemn, and solid: acorns, fruit, birds pecking at the fruit. But the bidding went way beyond me.

The hazel buds are well burst and exploded now. Even more like beetles in flight than they were before. The tiny leaves are expanding – they are expanding and the bud-cases are not. And I recalled as I looked that the expansion of these leaves from this stage on, from tiny flakes to final full size, is mostly the product of the expansion rather than the division of cells. It's the same with the thale-cress plant. When a leaf emerges as a bulge of cells on the flank of the meristem, that first phase of growth involves both the production of new cells by division and the expansion of those cells. But in the second phase, the phase that is in fact the major phase of leaf expansion, the production of new cells is less of a factor. It is the expansion of cells that is then the major contributor to leaf growth. Plant cells are very different from animal cells in this regard – they drive the growth of the leaf by their capacity to expand to volumes that are much greater than where they started. Animal cells tend to drive growth by making more cells.

Thursday 25th March

ON THE PRODUCTION OF FLORAL MERISTEMS

The sense that seeing and thinking and feeling all interact, add up to our perception of the world. And that perception is what the science is all about.

Looking out of the office window at John Innes, watching a procession of slow-moving clouds in a blue sky. Despite the slow pace, the yellow light not steady but booming full on like an opening flower, only to fade then boom again. Thinking again about the thale-cress in St Mary's. Has it begun to flower? Of course, even if it has, there still will be no flowers to see. But if it has, the meristem at the tip of the shoot will have been transformed.

The inflorescence meristem that a vegetative meristem becomes makes bulges destined to become another kind of meristem, a floral meristem: a ball of cells from which the sepals, petals, stamens, and carpels of the flower will be derived. Three different kinds of meristem successively build the shoot of a plant: vegetative, inflorescence, floral. The thale-cress plant contains genes that, when they are activated, tell the floral meristems that they are different from the inflorescence meristem that made them. Two genes, called *APETALLA1* and *CAULIFLOWER*. If these genes fail to work, structures that should become floral meristems remain as inflorescence meristems. Mutant plants that lack *APETALLA1* and *CAULIFLOWER* activity are bizarrely attractive. In such plants the inflorescence meristem makes a spiral of inflorescence meristems (rather than floral meristems). Each inflorescence meristem itself makes a further spiral of inflorescence meristems, and each of *these* inflorescence meristems in turn makes yet another spiral. And so it goes on. On and on in a great proliferation of meristematic tissue, with what should develop as a flower being replaced by something that looks like the head of a cauliflower. And indeed it *is* precisely akin to the head of a cauliflower, because a real cauliflower plant contains a mutant form of the cauliflower version of the thale-cress *APETALLA1* and *CAULIFLOWER* genes. *APETALLA1* and *CAULIFLOWER* encode transcription factors that switch on the particular pattern of gene activities needed to make the products of the inflorescence meristem into a floral meristem. A simple but profound alteration of identity that occurs during the development of most flowering plants but that fails to happen in the mesmerically ramified growth of a head of cauliflower.

The *Brassicaceae*, the family of plants to which the cauliflower belongs, exhibit remarkable plasticity of form. Cabbage, broccoli, brussel sprout, kohlrabi, and even thale-cress itself, display a fantastic range of shapes and structures. Yet, at the level of the genes, they are all much the same as one another. What distinguishes them is a handful of genes that have different activities. It may be that further changes in the activity of genes, small, incremental changes, gene by gene, over the course of millions of years, separated the cauliflower from ourselves.

Friday 26th March

Cold and bright, patchy light. The view of clouds from my window is picturesque, romantic, leads the mind in fantasies. Actually I think I see the beauty of it all differently since seeing the lovely things in Banningham earlier in the week. I see such landscapes and forms. Shapes. Topologies. Rocky outcrops. Misty wisps. A balloon hovering in the far distance. And there is something so extraordinary about the clouds' seeming solidity when in reality they are just vapour. Their colours too are so varied: shining cream against the blue sky, elsewhere a broad brush-stroke of lead grey. A world of sunshine and shadow. Seeing things in this way can only help with my current lack of vision.

This morning I start sketching a presentation I have to make to my colleagues in a few weeks' time: an overview of current and future research plans. It's a useful exercise – and might help me to get unstuck.

Tuesday 30th March

The clocks went forward at the weekend – and suddenly it's as if we live in a world of light rather than dark. Such a release. Spring is here, and with it a sense of relief – of healing.

The awareness of progression is strong. A procession that is accelerating. In the garden – celandine, borage, hyacinths, comfrey; all in flower. Buds on the sycamore turned to pale green and soon to burst. Those of oak, lime and horse chestnut close as well. The beech hedge will shortly shed its drab brown winter leaves and replace them with new ones. I'm so happy to be in the midst of it all, the hardness softening.

The hazel leaves are now expanded to a size larger than that of the bud scales that so recently covered them. The leaves exquisite miniatures of their future selves. Ridged and veined and with serrated edges. The tips of the serrations coincident with the terminus of a vein. And such an iridescent green, a green I can scarcely describe, that writing 'green' does not express, but that I do not know how to qualify.

Yesterday, I reread some of what I've written here. I thought I'd see if I could identify threads, continuing themes. I think that some are clear: the advance of the seasons in garden, fen, and wood; the growth of the thale-cress plant as part of that advance; a record of events in the lab; an account of our research, of its deepening understanding of the hidden mysteries of growth.

Yet I find myself stumbling as I write. The word *research* an impediment. Discomfort with the idea that I 'study' something, that that something is 'biology'. Why do I feel like this? Why do these words make me pause?

Is it that they are isolating terms? That I 'study' the growth of the hazel leaf, or the activity of a gene, by isolating it from the rest of the world? That by doing so I cut it off, reduce the connectivity of one thing with another?

APRIL

Friday 2nd April

YESTERDAY I heard that another of our papers has at last been accepted for publication. Unlike the previous one, this has been a relatively long, drawn-out affair. The first version heavily criticised by the reviewers (some of it fair criticism). Now our revised version is accepted. But there's a sting amongst the plaudits. One reviewer is still doubtful – says we're 'taking a risk' in publishing this paper.

This kind of comment unsettles me. We are under such intense pressure to publish. There is always the underlying concern that we might be publishing too early, before everything is checked absolutely, and that we get something wrong. And I'm frustrated because I still can't see where to go next. Perhaps I'm trying too hard. Keeping my bloodhound's nose sniffing on the trail, blind to anything but the scent. Blind to things that might help when the scent has gone cold.

Sunday 4th April – Palm Sunday

To Wheatfen. Light brilliant. Sky a moving picture of cotton-ball clumps of cloud. Wind strong, easterly, cold.

Yesterday, we watched the Grand National on the television. I loved it, the life-in-death sense of it, the rush, the haste, the winning jockey making the sign of the cross on his chest. We had a sweepstake, and Jack almost won. His horse fell at the penultimate fence. Huge tantrums the result.

And today it is Palm Sunday. The year moves on.

Out into the expanse of the fen. Last year's growth now looking dirty and tired, battered by the winter. Reed stems fragile and breaking at the nodes, wispy-thin. Mud-buff-brown. But, at ground level, the green is gaining on the brown. New leaves of grass, new blades of reeds. Leaves made in a manner both similar to and different from the way the leaves are made on the thale-cress plant. Similar in that they begin as a bulge of cells that appears on the flank of the meristem – but different in that they're pushed out in long parallel lines of cell division and expansion, unlike the more complex orientations of expansion and division that make the rounded leaves of thale-cress.

The spring is now most obviously advancing. Everywhere buds are bursting – on the hawthorn with pointed leaf clusters; on the willow beneath which I sit. But the ash buds are still black and closed.

The heads of the bulrushes now look completely different. Last time I was here they were firm and hard, dark chocolate. Now they're a ball of wool-white floss on a stalk – a tangled mass of seeds and hairs flapping in the wind. One stands tall above me. I pull it down to my height, bending the stem, and as I reach up to it I briefly see the woolly club against the blue of the sky as though it is one with the fluff of the clouds.

But the wind penetrates and I walk to the wood for shelter. On the way, I stop off at the broad. Here it's a little more sheltered, and there is a harmony of colours and textures. A pointillist fuzz from the distant opening catkins and willow flowers, a yellowness adding to the silver-grey haze of twigs and opening leaves, and consonant with the buff of the reeds below. On the water, a pair of swans. And a pair of little grebes floating on and diving through a surface that shivers and occasionally has depressions flying across it from the increased pressure of the gusting wind. Just now a butterfly – but gone too quickly for me to identify it. And above everything the endless blue sky, the clouds in stately procession.

In the wood such a feeling of shelter, of security. The birdsong all polyphony, the waves of hiss and roar from the wind in the treetops above, whilst on the ground I feel only a gentle breeze. There's a thrill of comfort in the distance of the roar – a sense of protection. I'm soothed. It's all so lovely and generates a feeling of joy, of transient but complete happiness. I love this place and time.

We're at a stage of the spring where, despite all the bursting and expanding green, the twiggy-ness of the trees is still their most obvious characteristic. But soon it will be different. Soon it will be all green-shady, the linear texture of the twigs softened.

Thursday 8th April

Weather volatile. Shifting at a fast pace. In the space of minutes an approaching cloud covers the sky. Making leaden light beneath it.

Hail slams the ground. The stones rounded, smooth-sculpted, imperfect spheres. Then, shortly afterwards, bright light. But the cold is constant.

Still no obvious sign of the thale-cress flowering. It is continuing to make leaves around its stem. Perhaps I was premature in reflecting a few days ago on the floral transition.

Sunday 11th April – Easter Sunday

To Yorkshire for Easter, to Starbotton in Wharfedale. Since Good Friday it has been cloudy and cold. The spring's progression at a stage behind where we left it in Norfolk. At first this morning the sky was still, unbroken, grey cloud. But then there began a slow explosion of light: in minutes the clouds parted and evaporated, the sun shone bright and clear.

Suddenly I remember how much I love this landscape. A landscape within which I spent part of my childhood. The way the light picks out, enhances in shadow the contour of the land. The grey rock and scree, the long, low levels of the fells – seen from the distance as rounded sculptures with dark behind them. I enjoy seeing these shapes and forms. They induce a sense of calm, of security. Could we ever think of the topological, the sculptural aspect of proteins with this same sense of comfortable familiarity?

There is more to the sense of security. The dale has the characteristic graceful U-shaped curve of a glacial valley. The basin of the dale a flat emerald plain of fertile fields. The dale sides sloping up to the moorland tops. There is a delicious tension between the nurturing meadows of the dale bottom and the wild roughness above. The feeling of security enhanced by the contrast. Akin to being indoors in a gale, in a port in a storm.

Monday 12th April

A wonderful walk from Starbotton to Kettlewell and back. The children scampering at our heels. First up the steep stony track to Cam Head. Then along the long, sloping green lane that descends into Kettlewell. Spectacular views down Wharfedale, and to our left the massive basking whaleback of Great Whernside, dead heather and rusty bracken brown in the spring sunshine. In Kettlewell, I found some thale-cress growing on a wall. Already in advanced flowering, all long, thin-stemmed, and white petals. Why so, when the Norfolk plant had not even begun to flower when last I looked at it? Natural variation perhaps?

Returned to Starbotton from Kettlewell along the bank of the Wharfe. The acid-sweet cheep of a wren bobbing in a bud-studded hawthorn.

Thursday 15th April

What I'm trying to write here is a world view that has science as one aspect. Rather than the way science is usually written – all microscope-focused with the ring around the lens enclosing/encompassing all that is seen.

And speaking of views, what wonderful things we've seen today! A day of bright sunshine and cool breeze. Walked the path that runs along the bank of the river – from Starbotton to Buckden to Hubberholme and beyond. Trees – sycamore and horse chestnut – springing into leaf. The soft shapes of the fells – Buckden Pike shining in the light.

The walking worked its magic on the mind. I found myself thinking about GAI. It was watching the shape of a cloud change over several minutes that did it, made me wonder about the form that the GAI protein takes in the cells of plants. Because strangely, whilst we know that the shape of the GAI molecule is vital to its function, that it flips from one alternative shape to another, and that this flipping is essential to its function, we do not know what either of these alternative shapes actually are. But by writing this I am ahead of myself with respect to the telling of the GAI story.

In the evening Alice and Jack giggled over a tape – Martin Jarvis reading *Just William*. At the end there is a fragment of a ragtime-style dance tune played on the piano in a way that's so expressive of the passing of time with the fall of each beat – gone and never to return. Such a sense of reckless-energy-flaunting-the-precariousness-of-our-existence.

Friday 16th April

Our last day here in Starbotton. Grey and cold again, drizzle and rain. Walked Starbotton–Buckden over the side of Buckden Pike, then back along the river-bank. Clouds not too low to spoil the view from above Buckden, a fine view of Upper Wharfedale up to Yockenthwaite and beyond.

Aware that this short time of rest – a separate existence almost – is all but over. But I'm ready to return to work. Perhaps a little apprehensive of that review of my work next week. But I know I'll be fine.

Sunday 18th April

We've returned to Norwich. A windy, grey, wet day. But there has been amazing progress in the spring in the ten days we were away. Oak and beech – even the limes – at various stages of bud-burst or young expanded leaf, flowering cherries in blossom, forget-me-nots profusely in flower. Everywhere a greenness and a softening of lines that was not there before.

Monday 19th April

A HYPOTHESIS

Back to the lab today. Preparing for tomorrow's review of our work.

Picking up the thread of the *GAI* story again, I'm going to write about what we discovered as a result of the cloning of *GAI*. About the 'relief of restraint' hypothesis. A step towards a unified theory of plant growth. Something that explains the spring leaves, the flowers in the garden.

There are two things here. There is what was seen: the results. And there is the vision, the enhanced seeing, that the mind creates from the raw thing that is seen. Although of course the one shades into the other so that it's hard to know where the one ends and the other begins.

So what did we see? What did the isolation of *GAI* reveal? There was the DNA sequence of the *GAI* gene. At first sight, a DNA sequence appears to be a prosaic thing. A string-sequence of four letters: stretching for thousands upon thousands of base pairs without obvious underlying form. But when a segment of DNA sequence is analysed using a computer larger-scale structural features become visible. Open reading frames, for example, the regions of DNA that encode proteins.

What distinguishes open reading frames? They are read in groups of three bases, each group being a 'codon' that represents one of the twenty amino acids that comprise a protein. For instance, the sequence of fifteen bases AGT TCT AGA AAC CTT encodes a polypeptide, a protein fragment, of five amino acids: serine serine arginine asparagine leucine. But some base triplets (TAG, for instance) are so-called 'stop codons'. Stop codons do not signify an amino acid and instead mark the end of a protein sequence. Since most proteins are upwards of 300 or more amino acids in length, and almost always begin with methionine (signified by ATG), a computer can scan a raw DNA sequence and identify likely regions of open reading frame. Such regions begin with ATG, continue with a section of 300 or more amino acid codons, and end with a stop codon.

It was by performing such an analysis that we first identified the *GAI* open reading frame, the part of the gene that encodes the GAI protein. Preceding the *GAI* open reading frame is the stretch of DNA that doesn't encode protein but that controls the activation of *GAI*: determines whether *GAI* is 'on' or 'off'. The promoter.

The *GAI* open reading frame encodes the chain of 532 amino acids that comprises the GAI protein. But a protein is a three-dimensional structure, not simply a linear chain. Its shape is the product of the folding and self-twining of that chain. The final structure is a function of the particular sequence of amino acids (each of which has different chemical properties) from which it is composed. The amino acid chain of GAI folds to make a unique shape. A molecular sculpture with an overall structure, specific surface features: crevices, pockets, protruding regions, the whole form relating to the molecular interactions in which GAI engages to perform its function. When trying to imagine this it might help to have in mind shapes of particular significance: a familiar landscape, a sculpture by Henry Moore . . .

Computers can also be used to identify particular regions of protein sequence. For instance, once a new sequence has been determined, that sequence can be compared with the many thousands of previously determined sequences. We did this with GAI and found that the final two-thirds of its sequence is closely related to that of another plant protein called SCARECROW (SCR). Since SCR is thought to be a transcription factor, a protein that controls the activity of other genes, it's likely that GAI is a transcription factor as well. We guessed that GAI controls plant growth by regulating the genes that encode the enzymes and structural proteins which are the real growth machine of the plant.

But it is the first third of the GAI protein sequence that is particularly interesting. This region has no detectable similarity to any previously determined protein sequence. It was completely new. Undefined territory. And our fascination with this region became even more focused when we obtained the DNA sequence of the *gai* gene.

To our intense interest, on comparing the DNA sequences of the *GAI* and *gai* genes, we found that the *gai* mutation affects that first third of the GAI protein. This finding was a source of great satisfaction, because it confirmed the very prediction that we'd made years previously: that *gai* encodes an altered protein rather than no protein. The *gai* mutant form of the gene differs from the *GAI* normal form by a deletion of sequence from the open reading frame – a small deletion that simply removes a segment of seventeen amino acids from within the first third of the encoded protein sequence. This removed region came to be known as the DELLA region – named after the first five missing amino acids (there's a single letter code for the amino acids) of this seventeen-amino-acid region. Here, to make it all clearer, is a sketch comparing the important features of the GAI and gai proteins (drawn linearly rather than as the folded structures they actually are).

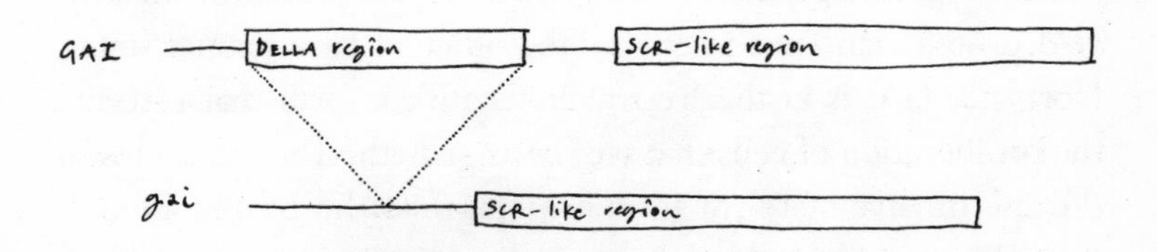

How the normal GAI protein differs from the mutant gai protein. Both proteins contain the SCR-like region, but the *gai* protein lacks the DELLA region (regions not drawn to scale).

So we had proved our prediction. The *gai* mutation does not destroy the gene. The *gai* mutant gene still encodes a protein, and that mutant protein still does something. But it has an altered structure because it lacks the DELLA region. This alteration must alter the properties of the protein, change the way it works. Ultimately, this change causes plants to grow dwarf rather than tall. A tiny change in the structure of a protein, invisible to the eye in the way that the protein itself is invisible, has a hugely visible consequence for the growth of a plant.

And then I saw my hypothesis. It was a process of thought that involved logic and something more. Like looking at a reed-bed in late summer when it's full and lush grey-green, with the stems of the reeds waving in the breeze. I cannot look at such a thing without making something out of what I see. Without seeing lines and shapes in it, connecting leaf to stem even when I cannot really see the junction. Penetrating the thicket of it with my vision. Relating the orientation of the parallel lines of the leaves to the prevailing direction of the breeze. Taking it just a little beyond what is actually seen, so that what is seen is transformed in the mind. Squashing shapes, pulling at them or adding to them to make a better fit or to form some new connection. These things are part of the generation of hypotheses, the stretching of the logic of 'if A is B then C'.

The hypothesis describes the growth of plants. Imagine that the GAI protein can exist in one or the other of two distinct states. Normally GAI is in the 'restraining state', a form that restrains the proliferation of cells that we see as growth. The other form is the 'permissive state', a form that permits the proliferation of cells. Imagine also that the growth-promoting hormone gibberellin promotes growth by causing the conversion of the restraining state of GAI to the permissive state. Previously, I described how mutants that lack gibberellin are dwarfed plants. If the above

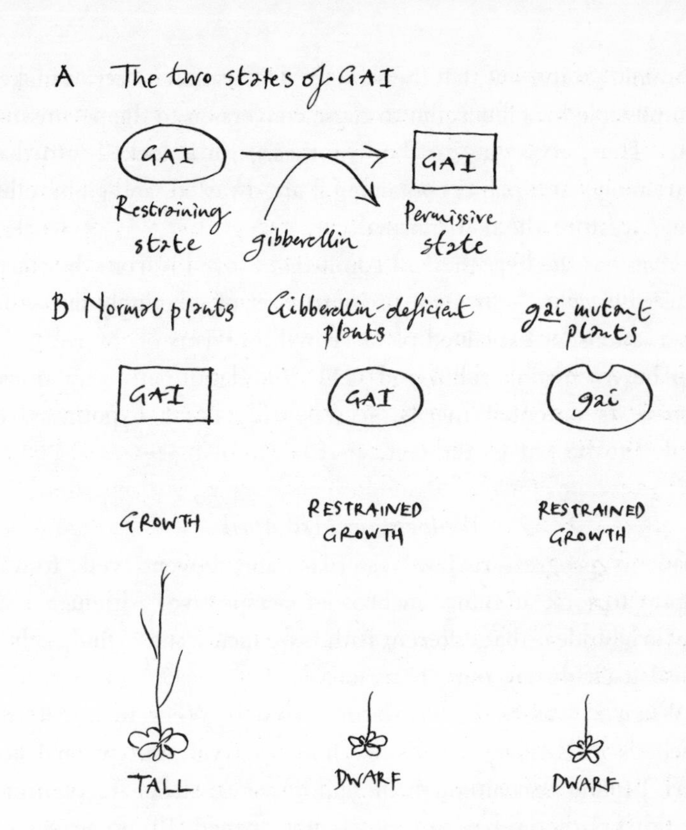

The hypothesis to explain how GAI works. A. The two states of GAI. B. Normal plants grow tall because gibberellin converts GAI into the growth-permissive state. Gibberellin-deficient plants are dwarfed because GAI remains in the growth-restraining state. *gai* mutant plants are dwarfed because the mutant gai protein cannot be converted to the permissive state.

hypothesis is true, gibberellin-deficient mutants are dwarfed because gibberellin-deficiency causes accumulation of the restraining state of GAI. Imagine further that the gai mutant protein still works as a growth-restraining protein, can adopt the restraining state, but that the structural alteration it carries makes it impossible for gibberellin to cause conversion to the permissive state. Thus, according to the hypothesis, gai is locked into that restraining state: plants containing it are dwarfed, and gibberellin cannot restore them to normal.

That was the hypothesis. I admit that I loved it from the start. It just felt right. It brought order to a set of seemingly disparate observations. It explained plant growth in terms of the relationship between gibberellin and GAI. But despite its attractions, despite its potential merits, it was still only a hypothesis. It needed to be put to the test.

Wednesday 21st April

I had my progress review yesterday, and it went well. It was helpful to look at things in broader perspective. Although that new bright idea, that different path I so much want to find, didn't reveal itself during our discussions.

When it was over, I rewarded myself. Went to see if the bluebells in Earlham woods are flowering yet. But when I got there I found that although the inflorescence stalks are well up, the flowers themselves are mostly not opened. I'll go again in a day or so.

And today the spring is so very lovely. The sky is a soft grey, but that only seems to enhance the lushness of everything, the sense of rush. The pace rapid over the last few days. After the cruel slow progress of February–March–early April, it's as if suddenly we're over some sort of threshold and now everything is easy – it's easy for the flowers to open, the buds to burst, the leaves to expand. All over the garden there hangs this sense of excitement with the expanding juicy greenness.

And there's delight in the knowing, the deep knowing that this expansion is a property of the protein we discovered. The GAI protein. That it is somehow a part of this beloved thing we call the spring. The trick is to keep the spring's beauty and the knowledge of the protein together in the mind at one and the same time. Not easy to do. I often lose it. But that trick enriches thought.

Thursday 22nd April

ON FOUR-DIMENSIONAL SPACE

I've had a chance to visit the thale-cress plant again. Prevented the last few days by my review, and before that we were away in Yorkshire. For some reason, my mind was all linear today, all dimensions, lines, and vectors. Fizzing, seeing connections. I was excited that I could get out. At one moment there was a line that ran from the top of my head to the sun, appearing and then evaporating from the surface of thought as I pedalled the bike to St Mary's.

And then, when I got to the plant, I saw that it too has begun to shoot upwards. At the very centre of the rosette of leaves there is the shortest of short stems. Just a quarter of an inch in length, perhaps less. On top of the stem, crowning it, a bunch of flower buds arranged in a spiral. There's been so much progress here in the week or so that I've been away. At last, at long last, the plant is flowering! The stem is fragile, but it thrusts.

And the lines were back in my mind. Rapid thoughts about this newly discovered stem, the roots penetrating the soil, the linear–cylindrical nature of both. The concentric cylinders of cells within which run the vessels, the xylem and phloem, the vasculature, the plumbing, of the plant. The vessels that carry water and nutrients, that feed the growing cells of shoot and root. These great vessels extend as the stem extends, adding more cells to themselves, expanding them. As they approach the tissues they feed, the vessels bifurcate, branch, and reticulate into tributaries of the main stream. Ultimately, a sponge network is formed, a network that at once infiltrates and forms a skeleton for the leaves. The end vessels are tightly intermeshed with the cells they service, the one embracing the other.

There is linear geometry in this embracing relationship. Let's define space according to a series of lines of dimension. One: a single extending line, a first dimension. Two: a line perpendicular to the first. Together these two lines define the second dimension of a flat, planar space. A third line for the third dimension, up and down, making the world of solid, three-dimensional objects with which we are familiar. Four. Four? Four: the solidity of the three-dimensional shape riddled through with channels like worm-holes in a piece of driftwood, its surface tunnelling into itself. This fourth dimension is the dimension within which the relationship between organisms and their vascular networks exists.

Vascular networks are thought to have fourth-dimensional properties because these properties conferred success on those organisms that first evolved them. Because the structure of vascular networks was shaped by the necessity for optimal connectivity between cells. Single-celled organisms are self-sufficient. They can do all the things they need to do to maintain their lives, exchanging or absorbing materials and energy with the world around them. However, when organisms became multicellular, cells began to specialise, to acquire different properties, to play particular roles. In plants, the cells of the root are specialised for absorbing water and nutrients from the soil, whilst the cells of leaves are specialised for photosynthesis, for trapping the energy of sunlight and turning it into food. Each of these groups of cells needs contact with the other. They are mutually dependent. In the long-distant past, the evolution of such dependency established a pressure, a necessity to develop the most efficient means of transporting resources from one part of an organism to another.

Multicellularity evolved independently in plants and animals, yet both have a vascular system. In animals, arteries, veins, and their tributaries carry blood through the body. These vessels subdivide into networks that are as tightly enmeshed with the cells of the animal body as the vascular system is in plants. In plants, the xylem and phloem act as conduits connecting one part to another. In both cases there is the supplier (vessel) and the supplied (the cell). Natural selection has forced the structure of these vascular networks so that they provide a maximum surface area interfacing supplier and supplied. Optimising the exchange of resources. Minimising the time and energy needed to transport resources throughout the organism.

The idea is that the shapes of our bodies are determined by these fourth-dimensional relationships in a way that unifies life from the very large to the very small. That tiny new thale-cress stem, pushing its way up from its rosette base into the sunshine, has a certain diameter. This diameter is related to the entire body mass of the plant, and this is true of all plants no matter what their size. Once the mass of a plant is known, the diameter of its stem can be predicted, using an equation: $D = kM^b$. This equation is simply saying that the stem diameter (D) is proportional to the plant body mass (M) raised to a power b, where k is a constant. It is the value of b that is crucial. And it seems that b $= \frac{3}{4}$, that this is the best fit with experimental observations and makes sense in terms of fractal (fourth-dimensional) geometry. That the 4 in the fraction $\frac{3}{4}$ is there because the geometrical nature of the vascular network that defines body shape is fourth-dimensional. And it is suggested that there are hundreds of other correlations in nature that relate such things as the structure of organisms, their rates of metabolism, their rates of growth, to body mass raised to ¾ . These are called scaling laws. So that although the forms of trees, plants, and animals seem so very different, they are on another level very similar to one another, and that similarity is an expression of these basic ¾ -power rules.

This theory is particularly attractive because it unifies. So much of modern biology is concerned with the nature of difference – how one species differs from another, or how a mutation in a gene can cause a difference between a normal and a mutant organism. Much of my own work depends on difference. The scaling-law theory is particularly refreshing because it provides a way of looking at what is common – the properties that make living organisms similar to one another. A way of seeing the entire living world as a joined-up thing, connected as it were into a single fabric. Linked by conduits and vasculature, vessels and tubing. A way of seeing that the differences between the parts of life, the species and individual organisms that comprise it, are tiny compared with the differences between the fabric and what lies outside it in the infinite silence of space.

But I seem to have got a little off course here. I need to emphasise that today marks an important event in the life of the thale-cress plant. That the stem I've been waiting so long to see has finally begun to emerge from the centre of the rosette. That there are flower buds spiralling around that expanding stem. That the identity of the central meristem has changed. It is now an inflorescence meristem, no longer a vegetative meristem. In that sense, although the flowers remain unopened, the plant has flowered.

Friday 23rd April

THINKING AT THE ATOMIC LEVEL

Sunshine, warm air, fresh green leaves everywhere. Spring seems secure now. Not brittle, like it was a few weeks ago. The chances of a snap back into winter receding. And with this comes a feeling of potential. Of release. The spring, summer, and early autumn stretch ahead far into the distance. Such a relief to be rid of winter. There can be no question that the seasons affect the state of the mind. That life seems easier now than it did just a month or so ago.

Flowers are everywhere now. The yellows of celandine, buttercup, dandelion. The thale-cress plant approaching flowering.

Today I stood looking down again at that plant. Growing amongst the gravel, its inch-long stem and flower buds quivering in the breeze. I thought how little I've so far considered what the plant is made of. Considered the nature of the dust from which it is constructed, its elemental composition.

It is perhaps an oversight that I have not already thought of the plant in this way. But then again, perhaps not. I'm coming to realise that there are many ways of seeing this plant and no particular order to them. That thinking of the plant can evoke countless resonances if I will but hear them.

Depicting the plant by means of atomic symbols makes it more clearly a part of the earth on which I stand. And of the wind that stirs its stem. Atoms then. That's the place to begin. Yet hard to describe. If only we could smell or touch, taste or see at that level. But then again, I'm no expert. And I suspect that the image of the atom that today's physicist would recognise is softer than the one I'm about to sketch. When I think of atoms I see them as a level of the structure of matter, themselves divisible into subatomic particles glued together by electrical charge and other forces. Atoms come in different species, depending on the number and variety of the subatomic particles from which they are made. According to their individual properties, different types of atom have different affinities for one another. Atoms of high mutual affinity can bond together by the sharing of electrical charge, and it is these bonds that hold atoms together and that make molecules.

The body of the thale-cress plant is made of molecules that are mostly built from only a few of the many different atomic types: hydrogen, oxygen, carbon, nitrogen, sulphur, phosphorus, and a few more. Water: built from two hydrogens and one oxygen atom (H_2O), is the most abundant molecule in the plant. Cellulose: the fibre from which the cell wall is made, is a chain of glucose-sugar molecules, each sugar a precise arrangement of carbon, hydrogen, and oxygen atoms. The other molecules in the plant, the proteins in the cytoplasm, the lipids in the membrane, the nucleic acids DNA and RNA – all are made from the bonding together of various atoms in the particular combinations and configurations that give each of the molecules their characteristic properties. The body of the plant itself has particular properties. These properties are the product, the outcome of the combination of properties of the molecules comprising the plant. The individual molecules having their own properties conferred by the atoms that comprise them. The properties of the atoms themselves governed by the subatomic particles from which they are built. Layer upon layer, from the visible to the invisible, the plant exists on all of these different levels. We would best see that plant if we could see *all* levels at once. But this is difficult to do.

Now this plant, its stem dancing in the breeze, isn't a static construction. Far from it. Its molecules are in constant flux, continually being made and destroyed within the cytoplasm. For instance, during photosynthesis, the plant uses the energy of sunlight to break water molecules apart. This liberates the energy that is then used to construct the molecules with which the plant builds itself. The plant cell's cytoplasm is a complex cauldron of reactions, of molecular destruction and construction.

The science of biochemistry has charted the massive complexity of this mix. Has shown how molecules change progressively from form A to B to C, by the stepwise formation or breakage of bonds, the addition or subtraction of atoms, by fusion with or separation from other molecules. Has also shown how enzymes (special proteins) catalyse these reactions, regulating the rate of flow along the pathways. Biochemistry has mapped the metabolism of the plant. But, for me, the conventional chemical representations of molecules, the dry chemical formulae, the three-dimensional models made from interconnected plastic balls of different colours, somehow fail to capture that dazzling range of structures and sculptures, the taste and smell of them, the frenetic energy with which they shoot along the branching pathways of molecular construction and deconstruction.

Saturday 24th April

DISCOVERING THALE-CRESS IN NORWICH

Went for a run this evening. Suddenly, I stopped short in the mews passage that runs behind College Road. For there, in the crack of the angle between cobbles and crumbling brick wall, I saw first one, then a whole colony of flowering thale-cress plants. Looking very different from the plant in St Mary's. Leaves less luxuriant, more purple, dry, and brittle. Much further advanced in their flowering. Presumably it's to do with security, with availability of resources. Here these plants are growing at great risk. Their existence precarious. Their environment inherently arid, making them dependent on immediate rain. Not much of what one might call soil to provide a reserve of moisture or for them to root into. The St Mary's plant has the benefit of a depth of soil, a better reserve. It can relax, doesn't need to hurry into flowering, can build a substantial body on which to feed its offspring. But here, these plants need to race to make seeds before the summer droughts kill them. It's funny that I only noticed these plants now. I've probably been running past them and their preceding generations for the last five years or so without seeing them. Studying that thale-cress plant in the wild is opening eyes and mind.

I continued to run. Into the searing pungency of a flowering currant bush, and a sudden evocation of memory. A dusty room, me sitting on a ragged red-upholstered armchair. A friend lighting a joint. An aromatic blue smoke filled the room. A scent like an autumn fire. Thoughts of childhood: security, tea-time, indoor warmth. He smiled and passed the joint to me, having taken several pulls at it first. I'd never done this before. But I was swimming in the pool of security and sucked the delicious smoke deep into my lungs.

I don't remember the beginning of the resultant change in my state of mind. But I do remember that some minutes after I'd taken the smoke into my lungs, I began to experience a strange kind of unease, and that we went for a walk around the college grounds. A pair of wood-doves flew side by side towards us across the wide, green, closely cropped lawn. I watched them above the abyss of anxiety that was beginning to open in my mind, and saw them shine with an intense, impossible beauty. They glowed with perfection. Their flight was at first straight and unswerving, but as they neared us they altered their paths, maintaining always the same distance from one another, in a curved trajectory that seemed graceful and elegantly focused on the achievement of a distant but precise destination. It was a moment that seemed as clear and perfect as the sounding of a bell over the crescendo of dischord that was developing in my mind.

My apprehension intensified. Walking somehow dampened it, so we left the college and paced along the roads outside. We went into a pub, but I found it no sanctuary. My mind screamed against the jostling crowd around the bar. I couldn't stand it, and instead broke out to continue my journey alone.

I walked out into the darkening night, finding that the faster I walked, the less the fear. And whilst that fear was boiling in one part of my mind, so the walking seemed to help another part that was looking on to itself. Watching the turbulence, seeing how one thought would be left unfinished before the next was clamouring for attention.

And there were things to be seen that were both beautiful and alarming. It seemed that the cars approaching through the dark had eyes, that they were searching for me with the light of their headlamps. Probing the blackness with their beams like those wondrous light-generating fishes that inhabit the ocean deep. Attempting to read my mind.

And yet the other part of my mind remained detached. A voice commentating on my state, saying that cannabis contains a gene encoding a protein that acts as an enzyme. That this enzyme promotes the synthesis of delta-9-tetrahydrocannabinol (THC). That THC had gone from my lungs, via my blood stream, to my brain. Its shape fitting like a key into the lock of a protein in my neurons, changing that protein's shape, and triggering an alteration in my state of mind.

After hours of walking, I came upon a particular quadrangle of intersecting streets. A square circuit I obsessively completed time after time, one point in the loop punctuated by a piercing sweet-sharp scent, the scent of the remembered flowering currant.

There was comfort in the repetition, and after a while my anxiety began to wane. I walked home and lay on my bed. Within minutes I'd drifted off into a longed-for sleep.

This experience was terrifying. Not to be repeated. But a moment of vision none the less. A segment of time that endures in the memory. And writing about it now I can see that there are affinities between what is seen in that extreme state and the kind of vision I'm grasping for now. In my continued search for a way to gain at least a glimpse of that much-wanted new direction.

Sunday 25th April

ON THE GROWTH OF STEMS

Today has been a heavenly day. Blustery white clouds in the sky. Wind blowing moist, mild air across the land. A surge of energy.

Cycled fast along the track to Surlingham. Eager to see the plant and the fen. Feeling strong and optimistic. Spring progressing, summer on the way.

First to the fen. The exchange of green for brown is advancing. Green spikes and needles are emerging from amongst the brown straw-weave of the reed-bed. And the sedges are beginning to flower – their edged stems topped by jet-black flowers: male above, female below. Some of them already gone over from black to yellow, with extruded anthers and pollen spilling out. The first butterflies are in the air: orange-tip, brimstone, peacock.

Then to St Mary's. And there it was. The thale-cress plant. With an erect bolting stem of several inches now. Nodding in the breeze. The top a tight spiral of still-unopened flower buds. Beneath the buds a length of stem, a leaf. In the crux of leaf and stem, more flower buds. Beneath the leaf, another length of stem that had its base in the rosette of leaves.

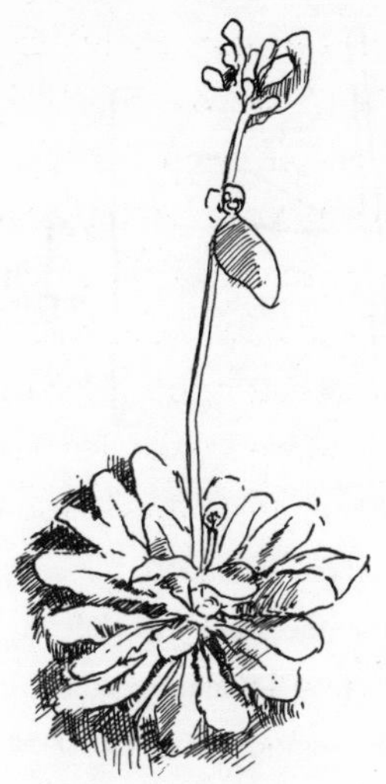

The bolting thale-cress plant.

These lengths of stem are segments sculpted from cells. As with the leaves and flowers, the cells that comprise each segment originate in the meristem at the stem tip. A group of cells on the flank of the meristem make an initial bulge. Then another group some distance round the flank make the next, continuing the spiral. The meristematic surface cells that lie between the two bulges will make the surface of the stem segment that separates those successive future leaves or flowers. The internal columns of cells that make the insides of the segment come from a group of cells at the base of the meristem.

I was marvelling today. At the uniqueness of that thale-cress stem. Since the beginning of time there has been no other like it, nor will there be another. Of course the cells co-operated in the building of this stem, in the way they always have and always will. But the precise construction of this particular stem is unique. A singular improvisation within the boundaries of constraint defined by a set of rules.

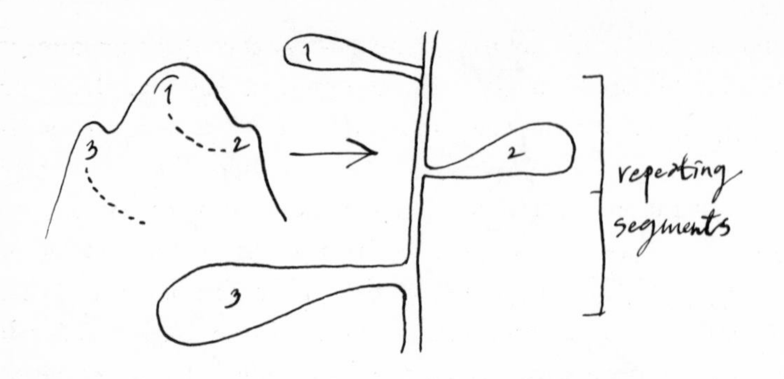

The growth of stem segments. First defined as the section between successive leaf primordia at the shoot tip (left; 1, 2, 3, etc.). The segments then grow to make a mature stem. In the vegetative rosette of thale-cress the segments remain short. The segments of the flowering (bolt) stem are longer.

Once the stem segment has been defined, it begins to grow. As first made in the meristem it is tiny, visible only through a microscope. Subsequently it expands into something that is clearly visible to the naked eye. During the past few days, the cells of the growing stem segments have been dividing and expanding, with rates accelerated by the warming weather. These rates are also regulated by an internal control, the GAI protein. We know this because the stem segments of gibberellin-deficient mutants are shorter, contain cells that are smaller and reduced in number when compared with the stem segments of normal plants. According to the 'relief of restraint' hypothesis, reduced gibberellin levels increase the growth restraint imposed by GAI. 'Relief of restraint' explains the development of a visible stem from its invisible meristematic precursor.

The thing that gives this particular stem its unique qualities is the fact that it grows by the proliferation of cells. A process that has a degree of predictability but an additional degree of randomness. As the stem segment-to-be leaves the meristem, it contains hundreds of cells. By the time it has completed its growth, it will contain many thousands more. There are mechanisms that control this process. That give an overall guide to rates of proliferation, restrict the orientation of divisions and expansions. But there is no single fate-map that directs the building of the segments. No precise pattern that tells each cell exactly what to do. It is similar in this respect to the growth of leaves. Each cell lineage, although partly constrained by 'rules', has a degree of freedom within that constraint. And if one considers individual freedoms for each of many thousands of cells, each making their own decisions within the constraining but not completely determinative context of 'rules', it rapidly becomes apparent that it's impossible for any one stem segment to be absolutely identical to any other in terms of its cellular architecture.

Looking at the stem of the plant I saw it in an overlapping way. Something familiar and predictable. But also, something singular and miraculous. The world is made of things that are at once expected and astonishing.

Tuesday 27th April

A NEW IDEA

The last few days have been simply lovely. Warm: 18–20 degrees centigrade, milky sunshine through a layer of haze. The weather generating a sense of pervasive contentment that is a context for all thought.

Today to Earlham woods to see the bluebells. They are quite magnificent. I sat on a log, an island surrounded by the blue and the faint bitterness of lily scent. The ocean stretching away from me in layers: the top layer of flowers blue-mauve; then a layer of green stems, parallel, perpendicular to the ground, erect save for the very top where they merge and curve into the region bearing flowers; then a bottom layer of leaves, glossy, shining green straps radiating out around the bases of the stems and licking at each other, squeaking at the pressure of touch or tred. And above all this the sky seen through the expanding leaves of beach and horse chestnut. Below, the earth: musty brown mould of last year's leaves.

The flowers: the different shades and strengths of blue and purple are exquisite. Not uniformly washed but with the intensity of colour structurally defined/limited: the petal underside has a line of deep purple up its central axis, with paler mauve flanks on either side. Each flower has six petals in a ring around six charcoal-grey anthers. The anthers borne on mauve stamens all surrounding a pistil that is pale purple at the rounded base, becomes stronger in colour along its length, especially strong in the stigma at the top. Such subtle and precise distribution of colour.

Amongst the multitude of blue bluebells there is one that is not distinguishable in shape or form from the others. Yet it is a brilliant white – stands out as an isolated bright point within the otherwise smooth texture of blue. It is a mutant. It carries an alteration in just one of its tens of thousands of genes. A tiny chance change in the DNA sequence of a gene that normally encodes an enzyme that is part of the metabolic pathway via which the characteristic blue/purple pigments are made. The mutant gene no longer works, the pigment is not made, the flowers are white in consequence.

Sitting rapt in the scene, surrounded by bluebells, I was aware of a happiness that will last, that will endure. That this was a moment I'd never forget. That the molecules of my mind would retain as a special thing. And there was also the awareness that molecules made the scene that was being retained: that GAI controls the growth of the bluebell leaves and petals; that enzymes make the pigments that give the flowers their colour; that those blue pigments are themselves molecules.

I had a new idea. About connecting the 'relief of restraint' model of plant growth to the 'tissue turgor' model – a model that envisages growth as driven by water pressure, the turgor or swelling of the cell. Will this idea develop? Who knows.

Wednesday 28th April

The sky dark. Heavy, steady rain. The ground drenched. Excellent for the growing stem of the thale-cress plant. But I can't see it today – I'm off to Southampton to do a seminar.

Friday 30th April

ON GROWTH FORCES

Back again to Surlingham. Last here on Tuesday. And there has been amazing growth in the time I've been away. The stem now a full half-inch taller than it was. Doubtless the growth was accelerated by the near-continuous rain of the past few days. The ground is soaked, the river high, the fen a swamp. The chestnut leaves are huge.

I wrote previously of the growth of the plant as the product of opposing forces. That there is the restraining force of GAI, and that GAI opposes a promoting force. But what is that promoting force? On one level, it can be seen as originating in the earth. The earth pushes the growth of the plant, generates a pressure that drives the expansion of stems and leaves. What, then, is the basis of this pressure? Where does it come from? It's actually a consequence of the properties of water, of its propensity to act as a solvent. The water within the cells of a plant is not pure water but is in the form of a solution: water in which additional substance is dissolved. At one level of abstraction, the cytoplasm of the cell can be seen as a concentrated solution of large molecules contained within the outer membrane that defines it. In contrast, the water in the soil is in relatively dilute solution. It is the disparity between these two solutions that generates pressure. Water molecules can pass through the membrane, but the larger molecules in solution within the cytoplasm cannot. The system inevitably tends towards equilibrium, to a state where there are solutions of equal concentration on either side of the membrane. Since the large molecules cannot move through the membrane out of the cell into the soil, the only way to approach equilibrium is for the water to move from the soil into the cell. The result: pressure.

This pressure is constantly pushing against the cell's walls. Its effects clearly visible to me this morning. Without it, the stem would not be standing so firm, so erect in the rain and wind. The cells of the plant are fat, turgid with water. As if the plant was a fountain, a visible form maintained by a pressure of water from under the earth.

This same pressure is the force that drives the plant's growth. On one level, it works like this: growth is the product of opposing forces. The restraining force of the walls of cells, the promoting force of the pressure of water from the soil. When growth begins, genes are activated. These genes encode enzymes that cause a weakening of the cell wall, a loosening of the chemical bonds with which the fibres that make the wall are meshed together. As the wall's restraint is weakened, so the cell's expansion is driven by the pressure of the water forced into the cell. And as the cell expands, it makes new material with which to patch the wall and strengthen it again.

Energy is classically understood to be the capacity to do work. So what is the primary source of the energy that fuels the work of growth? There is work in generating the water pressure that pushes the growth, and work in the construction of the new materials of the wall. The energy for both of these forms of work comes from the sun. Light drives the metabolism of the cell, and that metabolism establishes the concentration gradient across the membrane, thus generating the pressure. The new wall materials are also the product of metabolism. Ultimately, growth is light.

The realisation of a few days ago, in the bluebell wood, was one of possible equivalence. GAI exerts a force of restraint on growth. Previously I've thought of GAI as one source of restraint and the cell wall as another. But what struck me in that moment of realisation was that perhaps these manifestations of force have aspects in common. Perhaps they are even one and the same thing. Should this be the basis for a new line of inquiry? Have I at last thought of something new?

MAY

Saturday 1st May

OUT TO the north Norfolk coast. To Holkham in milky sunlight. The lumpy texture of the different trees, with their swelling and expanding leaves, some as buds, some opening, some flat, all stages in between: so variable. At Holkham that wonderful expanse of sand, the sea a blue rim.

Coming home in the car, moonlight shining on the road. Compare the fertility of the earth with the aridity from which that light is reflected. And further, 95 per cent of the universe composed of dark things: dark matter, dark energy. Utterly different from the stuff of which we are composed. So alien that its nature is simply unknown.

Sunday 2nd May

ON THE UNIVERSALITY OF GAI FUNCTION

As noted before, the *gai* mutant is very different from normal thale-cress. Dwarfed, darker in colour, the green more intense. The dwarfism not corrected by applying gibberellin. And the *gai* mutant also accumulates gibberellin to higher levels than normal plants do. This latter because gibberellin regulates its own levels. Since *gai* doesn't respond to gibberellin, this regulation no longer works, and gibberellin levels rise. Finally, the *gai* mutation is a dominant mutation.

But there was an important question hovering over our preoccupation with *gai*. Was the thing with which we were so fascinated simply a peculiarity? Something unique to the growth of thale-cress plants, something remarkable in itself, but concerning just one species amongst millions of others? Or did our observations reflect something more fundamental, something common to all?

We needed to answer these questions. One relevant factor was the existence in other species of mutants having similar properties to those of *gai*. For instance, the maize *D8* mutant is dwarfed, dark green. A foot tall versus the towering six or more feet of normal maize plants. Its leaves broad daggers, not long, thin swords. This mutant looks like other maize mutants that lack gibberellin, yet gibberellin applications do not restore it to normality. Furthermore, *D8* mutants contain high levels of gibberellin. Finally, the *D8* mutation is genetically dominant. In addition, there are mutants of wheat, the *Rht* mutants, that exhibit the very same spectrum of properties. These parallels with *gai* were tantalising. Pregnant with possibility. Perhaps they indicated the connection we sought.

Whilst maize and wheat are closely related to one another, thale-cress is only distantly related to both of them. Could it really be that the growth of such very different plants is regulated by something that is common to them all? Could that something be GAI?

The time when we were thinking about this was also the time when a number of different plant-DNA sequencing projects were in progress. We already knew, from the comparison of GAI and gai, that the DELLA region was crucial to the biology of GAI. So we looked to see if a sequence encoding a DELLA region (or something related to it) could be found amongst the then extant plant-DNA sequence data from plants other than thale-cress.

To our huge excitement, we found something. A stimulant to the imagination. A sequence from rice encoding something closely related to the DELLA region of GAI. Although *gai*-like mutants were not known in rice, rice is closely related to maize and wheat, and this new sequence provided a route to the isolation of *GAI-like* genes from them both.

But finding *GAI-like* genes in species other than thale-cress didn't really answer our initial question. Fundamentally, that question had been about function, rather than about whether or not such genes existed. However, the finding of these *GAI-like* genes put us in a position where we could now address the function question. That question could now be reframed like this: do the maize and wheat *GAI-like* genes control growth in the same sort of way that GAI regulates the growth of thale-cress? In particular, do the *D8* and *Rht* mutants carry mutations in the *GAI-like* genes they contain (in the same way that the *gai* mutant carries a mutation in the *GAI* gene)?

It was here that we met with huge difficulty. Sometimes science is like this. Technical problems that block progression. Cause mind-numbing frustration. When something that ought to be simple cannot be done. Once the sequence of a segment of DNA (like the *GAI-like* genes of wheat and maize) is known, it is theoretically possible to amplify that sequence directly from genomic DNA by a process known as the polymerase chain reaction (PCR). With DNA from thale-cress plants, this process usually works with little difficulty. But PCR can be affected by the nature of the sequence that is being amplified. For some reason, the *GAI-like* genes of wheat could not be amplified. Yet we had to be able to do this, because we needed to compare the sequences of the *GAI-like* genes from normal and *Rht*-mutant wheat plants.

For months we struggled with this problem. Wrestled with it. Tried all kinds of variation in technique. And it was particularly hard because there was no obvious rational approach. It became like cookery. Changing a recipe in blind hope. And then, one day, it worked. In desperation, we had tried a modification of the technique designed to help with the amplification of long fragments of DNA. The fragments we were attempting to amplify were not particularly long, and there was no real reason to hope that this modification would help. But it did.

At long last we were able to determine the sequence of the *GAI-like* genes from the *D8* and *Rht* mutants. And the results were so very exciting. The *GAI-like* genes from these mutants were of mutant form. The gene in *Rht* wheat was different from the gene in normal wheat. The gene in *D8* maize was different from the gene in normal maize. We could conclude that, just as we had previously found in *gai*-mutant thale-cress, mutation of the *GAI-like* gene had caused the characteristic dwarfism of the *D8* and *Rht* mutants.

We had answered the question. We had shown that proteins very similar to GAI control the growth of maize and wheat. From here on we can extrapolate. Propose that the growth of all plants is controlled in this way. From the iconic thale-cress in St Mary's, to the chestnut trees that surround it, to the reeds of the fen. Beyond the North Sea, in continental Europe, Asia,

America, the landscape of the entire world shaped by the activity of GAI.

There was another reason why this new development was so compelling. Previously, we'd shown that the mutant gai protein restrains the growth of plants because it's structurally different from the GAI protein. Because it lacks the particular sequence that comprises the DELLA region. Now we'd revealed a remarkable parallel between the mutation found in the *gai* gene and those found in the *GAI-like* genes in the *D8* and *Rht* mutants. Each of these mutant *GAI-like* genes encode altered proteins, and these alterations occur in more or less the same place as in gai. In essence, the mutant proteins encoded by the *D8* and *Rht* mutant genes, like gai, have abnormal DELLA regions. According to the 'relief of restraint' hypothesis, alterations in the DELLA region cause these proteins to restrain growth and to be resistant to the opposing effects of gibberellin. Because these proteins are all essentially the same, we renamed them after the something that is common to them all. The DELLA region. We named them DELLA-proteins, or DELLAs for short.

DELLAs affect our everyday lives. Most modern wheat varieties carry *Rht*, the mutant gene that encodes a dwarfing DELLA protein. The yields of grain from these dwarfed varieties are higher than those of normal taller varieties. *Rht*-containing wheat varieties were first developed in Mexico following the end of the Second World War. Subsequently the use of these varieties swept the globe. World grain yields rose. For some, starvation was alleviated. Who can tell what our world would look like now, what political or social changes there would have been, if wheat containing mutant DELLA proteins had not been so adopted?

But it was with the question of universality that I began this entry, and I'll end with it too. We had found that DELLAs control the growth of wheat and maize. This suggested that they are fundamental to the control of plant growth in general, and that we were not expending our efforts on something peripheral or idiosyncratic, something that concerned thale-cress alone.

Tuesday 4th May

Since Sunday it's been cold and wet, with periodic showers. Heavy rain bouncing on the ground. And today it's still very wet. The rain falls continuously, water penetrating my clothes as I cycle to work. And it's not mild or soft any longer. It's hard, cold, the sense of enlivening energy gone. A backwards step.

The cold got to me in the night. And I felt as I often do nowadays when it catches me unawares. I woke with a feeling of tension. Heart thumping and bumping, dry mouth, sensation of nausea. Unsure if the problem was that I was too hot or too cold, my mind too sluggish to know. But eventually I realised that of course I was too cold and went to boost the heating. Then waited for it all to come right. So strange that I feel the pressure but don't perceive it as being cold – although that's what it is. During the day, with mind alert – then I'd know.

Sensitivity is an important part of life. I respond to the environment, change my behaviour according to its changes. But the thale-cress plant is perhaps more sensitive – to light/dark, temperature, drought, and so on. More intimately connected to the world, because its position is fixed, it cannot escape adversity.

Lying awake last night I realised that it is some time since I wrote a proper description of what the plant now looks like to the innocent eye. That I should do this because there's been so much change, such progress, since last I did it. But I can't do it today. Too busy with meetings, admin, management etc., the paraphernalia of science.

And I've gone cold on my new idea. Can't see how to address it experimentally. So it's back to thinking again.

Thursday 6th May

THE BEGINNINGS OF AXILLARY BRANCHES

The plants have responded with such energy to the last few days of rain. The sheer mass of green so invigorating to eye and mind. So completely wonderful. But at the same time as writing this I am aware of discomfort in my mind. That although wonder is a[illegible] appropriate response, there is something that inhibits the ex[illegible] pression of it. Nevertheless, I do feel it. I will write this way[illegible] express fully the joy.

To St Mary's. The horse chestnut trees around the graveyard are regal, so splendid now. Only a few weeks ago they were skeletal, bundles of sticks that one could so easily see through. Now they are bells of green volume. Velvet, sensual-smooth, young flesh. The leaves close to their full expanse now. To this from the point of the bud in just a week or so – it's astonishing.

And so to the plant. Not perhaps so imposing, certainly not as massive as the chestnut trees. But still significant in terms of what it represents. Itself, the trees, all plants. The stem has grown a little more since last I saw it. Yet still the plant isn't flowering. In the sense that the flowers are not open and remain as closed buds. And new stems are beginning to grow out from the axillary buds that lie in the junction between the main stem and the leaves of the rosette. A crown of flower buds at the end of each.

Saturday 8th May

ANOTHER CRISIS

I think I knew that something like this would happen sooner or later.

To begin with, spring has in the last few days bounced back to the edge of winter. Today was a day of gusts, of parcels of brilliant light and the deep dark of fat clouds, of sudden rain. Despite the halt in progress, I do enjoy such days. The prick and sting, the energy, the suspense. That enjoyment fused with pleasure in thinking about the growth of the thale-cress plant, imagining its expanding stem. And I was thinking about all this on my way to St Mary's; about the water being everywhere: in the sodden earth, the wet sky and air, as well as in my plant. That water is the medium that connects the plant to the earth. I splashed along the track, sheltered briefly from a shower beneath a hawthorn. A few minutes of Lent, of drab light and chill and large raindrops. The shower passed and I cycled on.

The shock came when I entered the graveyard and walked to the grave. The plant had been eaten. The stem severed close to the base. A browning stump remaining, topped by a tiny bead of fluid. The leaves mostly gone, save for a remaining pie-slice-shaped segment of the rosette.

The damaged thale-cress plant.

It was a blow. The plant decapitated and largely destroyed. Taken by another life. Perhaps a rabbit? To think of it: some mangy, flea-infected, sore-riddled rabbit devouring my plant in the space of a few seconds.

And there's more. The other two thale-cress plants growing in the corner of the grave had been eaten as well. One completely gone, the other damaged beyond recognition but not so severely as my own special plant. Last week I had noticed thin green wires of grass beginning to carpet a portion of the grave near the thale-cress plants. These too are now cropped short.

Where is it now?, I found myself thinking in my frustration. Where is that rabbit? Doubtless somewhere close by, sleeping in its burrow or hiding in a hedge. In a day or so it will dispose of the remains of the plant. Pass a small, round, brown, mucus-covered pellet through its rectum and leave it on the ground. So that's it then. Project over.

After first finding the scene, my mind took unexpected paths. Looked at the remaining plant, the one that's less damaged. Perhaps I might transfer my attention, follow this survivor in my study? But I'm reluctant to do this. It seems wrong. Against the spirit of the endeavour.

A cloud drew near. Towering, menacing above the chestnuts at the churchyard edge. Iridescent purple on its top, brilliant orange where it merged with the sunlight. Beneath it fell clumps of hail that I saw as a fuzz of swarming bees, except that the fuzz had a fine grain of downward lines to it as though the bees knew the target to sting. There was a sudden flash of lightning and an immediate crash of thunder. Blinded by the light I looked down from sky to ground. Found myself wondering at the point of it all. What is it all about? One living thing devouring another, only to be eaten itself or finally to disintegrate back into the earth.

Then came the hail, stinging my neck. The white stones bouncing all around the remains of the plant, making it quiver with the force of successive blows. A further insult.

After the hail, soaking rain. Standing without shelter in the churchyard I was wet within minutes, clothes cold and clinging to my skin. Suddenly afraid. Frightened of death. I haven't thought about it much before. I've been through life denying death's inevitability. Perhaps it needed a moment like this one to force my thoughts to straighten up. How do we live with this?

Sunday 9th May

A PROTECTIVE CONSTRUCTION

Woke very early, when it was still dark. And in the hope of some salving of mind, I went out to Wheatfen in the gathering light to listen to the birdsong. The weather calm now, yesterday's storms gone, but still cold.

There is so much water in the fen – the level high in the broad and in the swampy part of the wood. On going into the wood, the sheer volume of sound filled my ears. An ecstasy of wrens. Tiny point-sources of noise dotted and bouncing around me in the thicket, strands of song bubbling and fluting. A fabric woven of pitch and rhythm, complementing the cloth of branches, twigs, and leaves. A pheasant's squawk (like the turn of a start-motor on a flat battery), a woodpecker's drum. And soaring momentarily above the whole, the high see-saw oscillation of the great tit. A blackbird's trilled arpeggios. Different strands woven in and out. The sound is captivating. The earth's chorus. The earth sings.

Looked later at the fen edge-on and it's like a two-toned bilayered ice cream. Top layer: brown. Bottom layer: sedge-rush-reed green. The green is pushing up, rising through the brown. What stately progress. What furious calm.

And then I went on to St Mary's. By now the air was warming. The sun well up, a feeling of expectancy. But when I arrived, I found further damage. The plant that yesterday was relatively unscathed is now completely gone. My own plant has only a few remaining leaves.

Despite my previous scruples, and although there seemed little hope in it, I intervened. I'd brought with me a length of chicken wire, used it to build a makeshift dome over the plant, edges held down with tent pegs. God knows what the people that tend the grave will think of this strange construction, should they return. Anyway, my hope is that it will keep the rabbits out.

Strangely, I'm more resigned now to what has happened. The rabbit (if it was a rabbit) no longer such a grotesque in my mind. After all, what I've witnessed is normal life. The life of the plant now the life of the rabbit. Each day our lives are sustained by death. The rabbit kills. We all kill to eat. Tomorrow we might eat the rabbit. This is an eternal cycle.

I knocked the tent pegs into the ground, their hooked ends pinning the perimeter of the dome of chicken wire tight to the ground. Of course a rabbit could still tunnel its way into this more protected landscape. But I think there is insufficient vegetation left in there now to make it worth the effort.

The sun was warming my back. I stood up from my work and glanced at the sky beyond the tower of the church. Approaching me was a storm cloud of slate grey, rain and hail falling from it in strong lines. And there were two huge complete arcs of rainbows, one in front of the other, so that the tower seemed in their very centre. Bows planted like croquet hoops in the ground and making a tunnel through which a vector could pass. The colours of these rainbows were of such an intensity, of such a violent purple in particular, as I've never seen in any rainbow before. I was awed by the sight of them, hope strengthened. I checked that the tent pegs were firm and returned to Norwich as the first rain started to fall.

Wednesday 12th May

ON SIGNALLING

Still grey. Cool and humid. But cycling between the University and the Institute I caught my first whiff this year of the sweet scent of the fen in summer. I don't know what it is – aromatic molecules in the air, of course – what I mean is that I don't know which plants make that particular cocktail of scents. But I do know that that scent is a stimulus received and the response an alteration in the texture of mind.

I'm thinking a lot about this at the moment, because of a conference I'm going to next week in Brittany. The subject of the conference: the relationships between signalling pathways in plants. So much of our thought at the moment is like this: there is a signal, the signal gets passed from link to link (entity to entity) of a chain that makes a linear pathway within the cell, until a response is elicited. It's like switch to wire to light. And the model works okay. There are many examples of pathways comprising such strings of components: protein A linked to protein B which alters the activity of protein C and so on. But I do wonder if the imagery of all this is so strong that it is forcing our thought, or way of seeing, squashing it into a shape that is such a distortion of reality that we're missing things. Example: perhaps the wholeness of the network of interacting pathways has a more substantial validity than the pathways themselves?

Thursday 13th May

There was something new about the world this morning. Something in the quality of the light. It flowed into the room when I pulled back the curtain, banana yellow and soft, something to swim in. Not hard, not vectorial like it is in winter. And the sleepy, moist warmth of the air was gentle and nurturing.

Yesterday, I thought about going to see the thale-cress plant. But the journey to Surlingham was too much for me to bother with when there seemed no point in it anyway. I was sure that it will soon be completely dead. Sometime soon, anyway. Today or tomorrow or the next day. So I didn't go. In any case I had so many things to do at work, to get sorted before going to France.

But today I went. I cycled along, by the river and over the track down into Woods End, shirtsleeves rolled up, arms bare to the sun and the breezes, aware of a yellow yellowness everywhere. In the track the lumped line of grass in the middle, on either side of that line the warming sandy parallels, and then on either side of them the yellow heads of dandelion and coltsfoot. Before Surlingham I went briefly to Wheatfen, and here were celandine and buttercups in a million flower points of sunlight, flies buzzing in the greening fen. I looked across to the flat horizon and the blue above it, and it seemed so unfair that, at this time so teeming with fertility, my one particular plant should have been killed. If only that rabbit had eaten something else.

I continued on to St Mary's. Knelt at the grave's edge. Began to lift the tent pegs, peel the chicken wire away from what it covered. The damaged leaves of the plant now dying back further with expanding patches of browning dryness. The green shrinking. I reached out to touch, to run my finger over the growing roughness. Then, lifting the edge of one scaly leaf with my fingernail, I found that underneath it there is another. A leaf that is green, alive, and undamaged. A leaf I hadn't previously noticed. I looked more closely. And there, to my surprise, I saw a small swelling crown of flower buds nestling in the niche that lies at the crux of leaf-stalk and stem. The domes of those buds felt as points on the skin of my fingertip.

There was a fractured second before the full recognition of what I was seeing. And then I realised. It's wonderful. A reprieve. For three days I've given this plant up for dead. But now there's a possibility that it will live. A fragile hope? Yes, but a hope none the less. For a moment I thought of resurrection. Although of course it had never really been dead.

The physiology of all this? Prior to the plant being all but destroyed, the shoot tip had been communicating with the rest of the shoot. Using a hormone known to plant biologists as auxin. Auxin is made in the tip, passes down the lengthening stem to the rosette of leaves at its base. Between the leaves of the rosette and the stem base to which they are attached, in the tiny corner where the one is attached to the other, lie dormant buds. Their growth restrained by the auxin. I'd not previously noticed that one of these dormant buds had been left undamaged by whatever had eaten the plant. Now that the source of auxin is gone, the previously restrained growth of that bud has accelerated. Bud and stem are beginning to grow. There's potential here. Already there are flowers (still only closed floral buds). In a few days, that stem will elongate and raise the flowers up from the drying rosette, a phoenix from the ashes.

Anyway, that is the hope. The question is whether this damaged plant has sufficient strength to support the growth of flowers and the making of seeds. Can it make sufficient food to keep itself alive?

I'm happy again. Happy with the hope this new growth brings. I re-covered the plant with the chicken wire and quickly cycled home. The whole journey in twenty minutes! Taking some tea leaves between finger and thumb, I dropped them into an old white mug. Poured boiling water on to them. Watched them swirl and settle, then still. As their motion slowed, the brown leaf cell-sap seeped, gradually pervading the water. I blew on the tea to cool it. My breath making transient flickering depressions, waves and troughs on the surface, whilst leaving the sub-aquatic world of infusion and rehydration that lay below undisturbed. Just as the delicate floor of a rock pool is unmoved by the sea breezes that play at its surface. And now I've drunk and sat for a while, looking at the leaves clinging via a meniscus of dark tea to the bottom of the mug. I can see the veins and conduits that once carried water, salt, and energy, that fed the cells. These vessels part of the continuity of branch, stem, and leaf that connect me through the tea to the sun. I feel its light warming my mind.

Saturday 15th May

It's 4.55 a.m. and I'm sitting in a train that's about to leave for London. From there on the Eurostar to the Gare du Nord, across Paris to the Gare Montparnasse, a TGV to Morlaix and then by bus to Roscoff in Brittany for the conference on 'cross-talk in plant signalling'.

All very exciting, and despite the initial sense of outrage when the alarm went off at 4.00 a.m. I'm well awake now, brain growing ever more zippy with the strengthening light. I think this is the pattern with me nowadays: mind peaking in fizz an hour or so after waking, the words welling up. Holding to that plateau of activity for a further few hours, then declining over the rest of the day.

Later. Hurtling through the green of Norfolk and Suffolk. Through hedges of white hawthorn, passing glimpses of yellow rape-fields.

11.30 a.m. (French time). Now on the Eurostar, speeding across the flats of northern France, towards Paris. Earlier, south of London on the England side, so struck by repeating images, one after and next to the other, of white flowering domes of chestnut trees, fading into the distant hazy horizon.

I'm wondering if on this journey there will be a moment of vision – when awareness combines with a sense that reality is more acutely felt. One of those moments that will endure – as when I was drinking tea a day or so ago. Everyone experiences these things. But they're fleeting and unpredictable. Sometimes I get them from my science. From an intense pleasure in the sudden realisation of the fit of things. Sometimes from the deep sense of peace that can pervade my mind when I'm travelling alone. As once, years ago, in the end carriage of a train crossing the fenlands of Cambridgeshire, facing backwards, looking along the lengths of the long, straight, parallel tracks as they tapered to a shimmering horizon. Best of all would be the new idea I need. But no such moments so far, though I'm cheerful enough.

14.05. Sitting on the TGV to Rennes/Brest. Four hours of peace – I can be alone with my thoughts – a pervasive sense of ease. So I'll doze a little, rest, write if I think of anything. Actually, all this train travelling is making me think again about lines, about paths and momentum and speed. About how science is sometimes conceived in terms of lines. There's a signal, the gibberellin hormone perhaps. It's recognised by an unknown receptor protein. Recognition activates the receptor. As a result, the receptor does something, modifies another protein (protein A; we've started: just seen the Eiffel Tower through a heat haze), and then the signal is passed on. This is a step: the passing of the signal from receptor to protein A (unknown), then protein A modifies protein B (also unknown), and protein B modifies protein C (unknown), and so it goes on, step by step, in unknown number until the signal reaches DELLA (now at great *vitesse* through lovely country only twenty minutes out of the Gare Montparnasse – woods, green wheat, horses in rough pasture). Modification of DELLA relieves the restraint that DELLA imposes and permits the growth of a plant, the visible response to the gibberellin signal. And the signal has velocity. It hurtles along the chain of steps in seconds.

France looks lovely through the carriage window. Wide meadows full of buttercups. Huge, empty provincial train stations baking in the sun. A sense of space.

Sunday 16th May

Looking down on to the little square in the centre of Roscoff from my hotel window. Architecture with style – wrought-iron balustrades, slatted shutters to the windows, frames edged with sandstone. A church tower from which the bell last night gave ictus and wow and stirred emotion/memory.

But right now I'm a little apprehensive. My talk to the conference this morning. A mixture of anxiety and excitement. But I think it will go well – my brain is sparking connections, and that usually makes it work.

Monday 17th May

The talk went well. My delivery was smooth and unstumbling. Confident (it's not always this way). Some interesting questions at the end. Then I sat back and listened to the other talks. Some beautiful pictures of thale-cress plants with leaves glistening in the light. Others of the hairs on the roots. Of course I found myself remembering the plant back in Surlingham. Wondering how it's doing. Is it managing to regenerate from that remaining segment of itself? Will it recover with sufficient vigour and body to flower and make seeds?

Outside, hazy sunshine, a cool, salty, seaweed-scented breeze. Views of rocky outcrops and islands in a flecked blue sea. And there we all were, shut inside. In a darkened room looking at our various plants and results projected on a screen, talking and discussing what it all might mean. And although I think that what we're doing is right, a necessary human activity, I do wish we could somehow connect more strongly what is going on inside that room with what is outside of it.

There's a graveyard just outside the conference building. It bakes in the sun. It's arid, not a blade of grass grows there. Just graves and gravel. So unlike the greenness of St Mary's.

Wednesday 19th May

On the train from Morlaix back to Paris. And I'm tired. Sapped. Poetry, luminacy, articulacy all shot. A dryness when the lids scrape closed over my eyes. Soon I'll fall asleep.

But it's been a good meeting. Nice to meet old friends. And I get increasingly convinced that DELLAs are key to growth, to many aspects of the growth of plants. So many things I've heard in the last few days point that way. Although in most cases these are possibilities, not robustly proven. But I do like the promise of coherence.

Somehow, today, tomorrow, over the next few days, I have to write a grant proposal to continue the funding of the research in my lab. I have to find a way to invest it with significance, to make DELLA glow. But now I don't have the mind for it. Indeed I'm hearing the old voice of frustration at the edge of my mind. Because I'm tired, I suppose. And still I can't see where to go. If I can't see where to go, how am I to write a grant proposal that is in any sense compelling? This whole 'lack of vision' problem is getting increasingly serious.

Thursday 20th May – Ascension Day

ON 'MASTER REGULATORS'

Returning home. This morning walked around Paris, watched it waking like an organism as the day progressed.

Then in the Jardin des Plantes. Warm sunshine. Tall, shady trees at the corner where the entrance is, spring greens in the leaves, birds singing in the branches. Further in, long avenues, geometrically sculpted and pruned in the French way.

Then a taxi ride to the Gare du Nord. There's music in the cab, and it's hot. Something sassy that has a top layer that pushes and pulls at the beat, fills in the gaps with baroque intricacy and tracery or with stabbing interjections. And all the time the beat is there and fundamental like a gong. The tension between that steady base and the fluidity above giving it all energy. I like the fundamental. DELLAs are fundamental in the same sort of way. But we should exercise caution in how we write these things. The way we name, provide terminology. Often things like DELLAs are called 'master regulators'. I'm uneasy about the resonance of this. Dislike the idea of 'control', the thought of masters and slaves. There's a discrepancy between the crude brutality, the ugliness, of such images and the elegance of the real thing. However, the 'master regulator' idea has currency. It has worked as a position from which to move forward.

Finally, I'm home. Its cool. Sunlight and passing stormy showers. From the garden there's a sense of fullness, of completeness. As though the spring expansions are all done. This, I suppose, is summer.

Friday 21st May

It has been warm whilst I've been away. But today there's a chill wind, and dark clouds that float with occasional menace.

This evening I took the children to Surlingham. Hoping that the plant would be okay. Last week it seemed that it was back on course. Up against it, yes; touch and go, for sure; sailing close to the wind, but still sailing.

Once in the graveyard, we pulled back the wire. And I was pleased to see that the plant is still growing. The new stem elongating, pushing flower buds up into the air. But it's a thin stem, not as thick as the original, a flimsy mast on which to hang a sail.

I was so happy to be back in the peace of the churchyard, amongst the cooing of the wood pigeons, surrounded by the stately chestnuts in velvet-green leaf (the white spires of their flowers nearly over now). To be recording the recovery of the thale-cress plant. In just a few days the cells of what was previously an inhibited axillary bud have become organised into a working inflorescence meristem. Segments of stem have been built and expanded. Flower buds made and pushed up into the breeze. As I looked at that fragile, narrow stem, a bumble-bee buzzed by, caught my attention, and took it in a curving path over the graves. I watched the bee until the flickering shine of its wings became lost amongst the glinting flint-mottled grey of the church tower. A moment of enhanced vision in which I saw proliferation.

The cells of the plants are proliferating from that remaining bud. Proliferating in ways that are restrained, that maintain the organisation of the organs they're building. What is amazing is that this growth is organised so well that an expanding stem-internode segment remains as such, does not become a swelling, unregulated spherical mass. Perhaps the fact that it's so commonplace, that it happens this way all the time all over the world, dulls the taste of it. But it's still something to wonder at. And suddenly I see my plant in a new way. As being built iteratively, of segments of a few kinds, stem internodes, leaves, sepals, petals, stamens, and carpels, connected in a particular order along the axes of its growth. Each of these different segments made by the restraining of proliferation in a way that shapes the particular segment they become.

But Alice and Jack were bored, clamouring for us to go on to Wheatfen. And in Wheatfen I saw it the same way. Amongst the scratching of the reed leaves rubbing against each other in the breeze. Within the flowering hawthorn of the hedge. Still I saw repeated segments. A world constructed of iterated blocks of defined identity. But also aware of the necessity not to make those blocks too rigid. That the blocks can warp in response to external things. Not one of them is absolutely identical to any other.

Alice started to play with the mare's tails. If ever there was a plant built from blocks, this is the one. It has stark beauty. Built from base to tip of segments that repeat themselves over and over, one placed on top of the next. At the joints between the blocks a circular fan, a stellar snowflake of smaller, thinner branches that grow out at an angle to the main stem. The whole thing expanding in the warmth and wetness of the fen.

I too played with them as a child. Rubbed my fingers against them, feeling how they were rough in one direction and smooth in another and wondering at it. Taking a stem and holding it between the fingers of each hand, base in one, tip in the other, and pulling in opposing directions, increasing the tension on the stem, until it suddenly snapped. Looking at the broken ends to see that the break was clean and flat, that there was no obvious damage, just a white smoothness on the new surface that the snapping had revealed.

The stem would come apart at the junction between two segments. I remember thinking that at that point perhaps there was a region of weakness, or of greater weakness than in the rest of the stem, so that as the tension of my pulling built up, so that point was the first to break.

And because I liked it I did it again. Enjoyed the uncertainty of when the stem would actually snap, the smoothness of the surface that the snapping revealed. I wanted to know if it would happen again the same as before. And then I'd do it again, and again and again, on and on until all the joints were broken, ending up with a heap of segments, all very like one another.

Then I'd push the segments together again, finding that the cleanness of the breaks made them fit tightly. When I looked up from my game I saw the leaves on the nearby trees as repeated units. Likewise, the petals in flowers. Looking back at the heap of segments I'd see that the branches that fan around the joints of the mare's tail stem are themselves built of linked units, and that each unit is itself a repeating structure. I'd see hierarchies of units, units within units, although of course I didn't know then how this rule extends itself inwards to the invisible parts, the cells and their contents, from which the stems and leaves are built.

There's a connecting memory, from years later, of landing at the airport at Los Angeles. It was dark, and as the plane descended, so a vast, diffusely shining landscape became transformed into a coalescence of contoured light. Then into individual lights, orange sodium, sulphur yellow, or bright white. Each light a tiny part of the whole. Often connected in strings: wire to light, wire to light. And, in the strange frame of mind that the long intercontinental flight seemed to generate within me, I saw our entire world as constructed in this way, of units that iterate endlessly, and that are all connected one with the other.

Monday 24th May

TESTING THE 'RELIEF OF RESTRAINT' HYPOTHESIS

A blanket of cloud plugs the sky's aperture. It's humid, mild. A sense of ease in life. Summer relaxation. And it's visceral. Muscles less taut, exerting less tension. Summer in the body.

To continue with the thoughts about segments, repeated units. The stem growing by their expansion. First defined in the meristem, then expanding. How do they expand? They expand because gibberellin overcomes the growth-restraining force of the DELLAs.

Thale-cress contains five distinct DELLAs. Distinct, though closely similar to one another. Previously I wrote about the cloning of *GAI*, and of how this gene encodes GAI, the first DELLA to be identified. Since then, the entire DNA sequence of the thale-cress genome has been determined. Thirty thousand genes (or thereabouts) revealed. Amongst these are four relevant ones: *RGA*, *RGL1*, *RGL2*, and *RGL3*. These particular genes have DNA sequences that are closely related to that of *GAI*. Encode proteins that are very much like GAI. Do these five proteins all control growth in the same way as GAI? That was the next question.

Actually, we knew about RGA before the completion of the genome sequence. Another lab had already shown that RGA regulates growth in response to gibberellin. But with the completion of the genome sequence we knew for the first time the full complexity of the family of proteins we were dealing with. When proteins have closely related amino-acid sequences, they adopt closely related three-dimensional shapes. Proteins of related shape usually have related function. Since all five DELLAs were so very similar to one another, it seemed likely that they would all act in similar ways to control the growth of a plant.

Now to the test of 'relief of restraint'. The DELLAs restrain growth. Gibberellin promotes growth by overcoming the growth-restraining effects of the DELLAs. The test of the hypothesis was dependent upon a prediction. The prediction was this: If DELLAs restrain growth, and gibberellin promotes growth because it overcomes the activity of the DELLAs, then a mutant lacking both DELLAs and gibberellin should grow tall and not be dwarfed. Put another way, a gibberellin-deficient plant is dwarfed because it lacks the gibberellin required to overcome the growth-restraining effects of the DELLAs. But if that plant lacks DELLAs as well as gibberellin, there's nothing to restrain growth. Such a plant should grow tall even though it is gibberellin-deficient.

The 'relief of restraint' hypothesis represented a new way of thinking about growth. For years the growth of plants had been known to be controlled by gibberellin. But mostly this had been thought of as being an active thing. That gibberellin simply promotes the growth of plants. The new hypothesis was suggesting something rather different, namely that gibberellin acts by neutralising a force that represses growth.

The hypothesis was not simple to test, because it considers DELLA activity as one thing. And yet we know that DELLA is not one thing but five things. There are five DELLAs. A complete test of the hypothesis would have required a mutant lacking all five DELLAs. Finding a mutant lacking one of them had been hard enough. Finding mutations that caused lack of each one of the five, and then crossing them together to end up with a plant line that was totally DELLA-deficient, seemed, at the time, to be close to impossible.

But we could at least attempt a partial test of the hypothesis. We already had a mutant that lacked GAI, another mutant that lacked RGA (kindly given to us by the other lab), and a mutant that lacked gibberellin. Perhaps it would be possible to test the hypothesis in a limited way, using these mutants. To ask if a plant lacking GAI and RGA (but still of course containing RGL1, RGL2, and RGL3), and also lacking gibberellin, was taller than a plant lacking gibberellin and having normal levels of GAI and RGA.

So we began the experiment. To understand the genetics, remember that plants have two copies of each of the genes involved. All of the plants in the experiment were gibberellin-deficient. They carried two copies of a mutant gene that causes gibberellin deficiency; these two mutant-gene copies are written *ga1-3/ga1-3*. The experiment also involved plants lacking GAI. These plants carried two copies of a mutant *GAI* gene (*gai-t6*) and two of the normal *RGA* gene (and so were written *gai-t6/gai-t6 RGA/RGA ga1-3/ga1-3*). Finally, the experiment involved plants lacking RGA. These plants carried two copies of the normal *GAI* gene and two of a mutant *RGA* gene (*rga-24*) (and so were written *GAI/GAI rga-24/rga-24 gal-3/gal-3*).

By performing a genetic cross, we aimed to make gibberellin-deficient plants lacking both GAI and RGA. We made a cross between a gibberellin-deficient plant lacking GAI (*gai-t6/gai-t6 RGA/RGA gal-3/gal-3*) and one lacking RGA (*GAI/GAI rga-24/rga-24 gal-3/gal-3*). To perform the cross, a pollen-shedding anther from the one was brushed against the stigma of the other. Then came the first wait. Had the cross succeeded? Would the seed pod of the pollinated flower bear fruit? Within a few days there was the pleasure of seeing the pod elongate, of watching its surface become bumpy with the swelling of the seeds inside.

We took the seeds made from this cross, gave them gibberellin (without which they won't germinate), let them germinate, washed away the gibberellin, planted them out. The seedlings that grew from these seeds looked just like gibberellin-deficient dwarf seedlings, like plants that carry two copies of *gal-3* and the normal complement of GAI and RGA. This is exactly how it should have been. This new generation of seedlings had inherited, for each gene, one copy from the mother plant, one from the father. Since mother and father each had two identical copies of *gal-3*, the only possibility was that the next generation would also have two copies of *gal-3* and would therefore be gibberellin-deficient. In the case of *GAI* and *RGA*, things were more complicated but meant that this new generation of plants would all have one copy of the normal and one copy of the mutant form of *GAI* and *RGA* (and would thus be *GAI/gai-t6 RGA/rga-24 gal-3/gal-3*). Because the normal form of the *GAI* and *RGA* genes is dominant over the *gai-t6* and *rga-24* mutants, these plants would be expected to look like gibberellin-deficient plants which had normal levels of GAI and RGA. As expected, these plants grew into dark green dwarf plants that needed gibberellin for normal flower fertility. We gave them the gibberellin, allowed them to self-pollinate, and watched the seeds filling in the seed pods.

The next generation of the experiment was the one that was really revealing. The plants that grew in this phase would be the ones that would provide the test of our prediction. The expectation was that one in every sixteen of them would lack gibberellin and would also lack both GAI and RGA (*gai-t6/gai-t6 rga-24/rga-24 gal-3/gal-3*). We took these seeds, treated them as before, planted them out, and waited. What would we see? Of course it was one of those things that starts off as 'I think I can see something.' Something on the edge of certainty, something tantalising because you've guessed that it might happen and now you think perhaps it has. And then gradually the 'I think' increases in confidence and becomes 'I'm sure.' As we watched the growth of those plants over the ensuing weeks, so we became increasingly certain that our predictions were being met.

Amongst these plants, at roughly the expected frequency, were some that were growing as tall as normal plants. Despite the fact that they were gibberellin-deficient. Tall and gibberellin-deficient? Surely a contradiction in terms. Under any other circumstances, these plants would have been dwarf. Yet we had predicted it. And further tests proved that our prediction was correct, that these tall, gibberellin-deficient plants lacked both GAI and RGA.

What excitement came with this! The hypothesis had met the test. The results of the experiment consistent with the prediction nested within it. 'Relief of restraint' is true. As always, this new discovery provoked a flood of new questions. Perhaps the

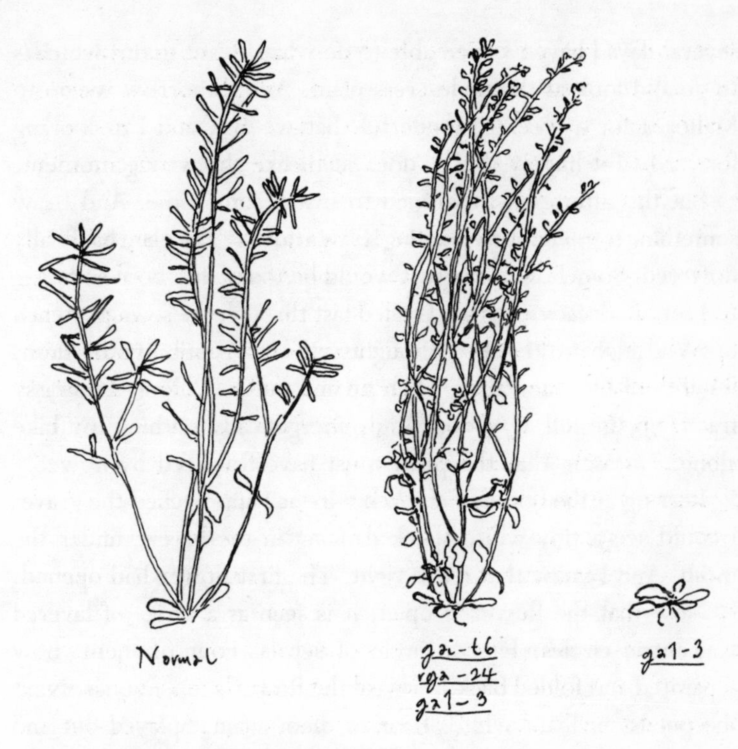

Comparison of a normal thale-cress plant, a gibberellin-deficient thale-cress plants lacking GAI and RGA (*gai-t6 rga-24 ga1-3*), and a gibberellin-deficient thale-cress plant (*ga1-3*).

most important of these was this: we had shown that gibberellin overcomes the effects of GAI and RGA. But we didn't know how. How does gibberellin overcome the effects of GAI and RGA?

Thursday 27th May

THE PLANT FLOWERS

The last few days have been frustrating. Endless meetings. Grant proposals to read, criticise, and write. Abstracts submitted for talks at the conference session I'm chairing in September. For several days I haven't been able to do what I want to do, which is to go and look at the thale-cress plant. And tomorrow we go to Mallorca for a week! Wonderful that we are, and I'm looking forward to it hugely, but it does seem like the wrong moment.

But this afternoon I managed to snatch some time. And I saw something tremendously exciting. I saw a flower. The plant has finally flowered. Somehow, I knew it would be today. It's been so hot.

I set off already excited. Cycled fast through the sewage-stench of Whitlingham. Evoked thoughts: the corruptibility of flesh, fragility of life, mortality. Then up and out of it along the grassy track, up the hill at Woods End, energetically pushing my bike along. Guessing that the plant must have flowered by now.

Peering at the dome of chicken wire as I approached the grave, I could see a tiny white speck dancing in the breeze under the mesh. And I knew that I was right. The first flower had opened.

Now that the flower is open, it is seen as a series of layered concentric circles. First a circle of sepals. Four of them, now separated and folded back to reveal the flower's inner zones. Next the petals, brilliant white. Four of them again. Splayed out and perpendicular from the centre. In the form of a cross. Inside again, the next layer. A ring of six erect stamens. Thin filaments topped

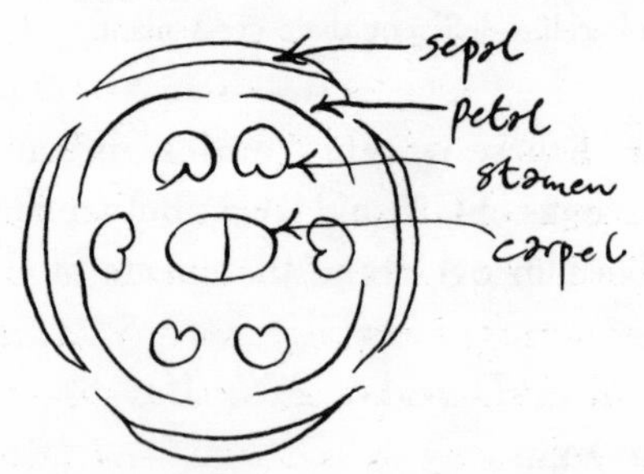

The layers of the flower. An idealised cross-section taken roughly halfway between top and bottom of the flower, showing the concentric circular whorls from which it is constructed: sepal, petal, stamen, carpel.

with yellow anthers, standing proud as pins. Finally, the carpels. Fused together to make a round pistil. Topped by a shaggy stigma.

This flower is a terminal structure. The final product of a succession of meristems. At the beginning, the shoot meristem made organs in spirals. Continuously, without any clearly defined end to the process. First, as vegetative meristem, it made a stem and spiral of leaves. Then, as inflorescence meristem, it made more stem around which it twined a spiral of floral meristems. And then it was destroyed.

But that destroyed meristem left the bud-meristem legacy. Every floral meristem contains an end point within itself. It is determinate and makes four closed rings, not a continuing spiral. Four whorls of organs in a defined order: first sepals, then petals, then stamens, then carpels. At that point, the making of organs ceases. So the flower is an ending, a culmination of the progression of meristems: vegetative, inflorescence, floral.

Vegetative → Inflorescence → Floral

The progression of meristems.

The mature flower grows from the floral meristem. As previously, the organs have their origins as bumps. As outgrowths of cells from the flanks of the meristematic dome. But this time those bumps are arranged in a ring, not as a spiral. Once formed, the bumps attain identities peculiar to themselves: sepal, petal, stamen, carpel. Following the attainment of identity, the organs grow to particular shapes and sizes. The whole flower enclosed within the sepals, the outermost of the rings.

And finally there is today's stage, the opening of the flower. It was still growing even as I was looking at it: the petals expanding, the stamen filaments elongating, the yellow anthers nearing the surface of the stigma.

It was a great moment. I began this journey back in February. Now, in late May, this flower has sharp significance. Something I thought I might never see. Of course I've enjoyed all of this year's flowers – the buttercups, daisies, celandine, and dandelions, the yellow flag-iris in the fen, so many others. But my pleasure in them was nothing to this. I suppose it's the continuity that does it, the sense of narrative. I've watched this plant grow from frost-scarred rosette, through several falterings, to final flowering. And there is another aspect to it. Whilst bobbing in the breeze, the flower connects with the earth through stem and root. Through leaves and veins to the sun. It's part of a whole, not separate.

I suddenly know, right now, what I'm about with this project. It's about seeing. Sharpening the focus of vision.

Sunday 30th May – Mallorca

On holiday. Yesterday I was all scratchy and irritable. Tired from the journey. Tired from trying to find the place where we are staying in heat that coils itself constraining as a python around action and thought. Noise gnawing like a sore – traffic, children, pneumatic drills, a jarring cacophony.

But this morning it's lovely. A swimming pool, the sunlight dancing on its floor. The pool set within an orange grove.

I've been thinking a lot about the naming of things. About how we scientists give things names or use turns of phrase, about the *language* of science and how it affects the way in which we think. I'll write more about this when my thoughts are clearer.

Just before leaving, on Thursday, I saw that our paper, the one I'd rushed with in January/February, is now published on-line. It looks good. Figures are clear and given enough space. It sums up, consolidates. A nice piece of science.

Monday 31st May – Mallorca

This place (Sóller) a pretty town. And there's no denying the beauty of the mountains that surround, the strength of the light, the vibrancy of the colours.

This morning I looked from our bedroom window out past a large palm tree, across the orange grove with glossy leaves shining in the light and globes of oranges dotted amongst the green, and was suddenly aware that no one had ever seen exactly what I was seeing at that moment, nor will I or anyone else ever see it again. But how to capture adequately the uniqueness of the moment?

JUNE

Tuesday 1st June – Mallorca

STRONG sunlight that hits hard at noon, despite the breeze. To some gardens a few miles away (the Jardins d'Alfàbia). Historically the house and grounds of the one-time sultan of this part of Mallorca. Built on the site of a spring high up in the mountains. It is laid out in sections – a cool, shady grove of palm trees, firs, and bamboo; a path lined with a tightly clipped hedge; a descent of stairs in an avenue of palms. Everywhere the sound and feel of water: bursting from the spring, dribbling, running and singing in channels laid into the ground, in ponds and fountains, rivulets and cascades. There's the sense that the water is the force that drives the garden. Especially when the children dance in a fountain.

The palms are so satisfying in their architectural, sculptural grace. Built by meristems. Those tiny balls of cells, maintaining themselves and building the tree whilst decades pass. The texture of the lines that originate in that ball: the line of the trunk, the lines of the palm fronds, and the lines of the leaflets that grow from them. Other lines all jumbled in the canopy above, chiming with the straightness of the sunbeams that fleck and penetrate through the tangle. The sharp divide between sunlight and shadow.

Wednesday 2nd June

Sky blue, haze, less hot. The scent of wood-smoke in the air this morning. It evokes thoughts of change, of transformation. Of progression, seasonal, autumnal almost, stubble-burning. And this is what scientific terms, for the most part, don't do. They lack resonance. Take GAI. An acronym. *G*ibberellic *A*cid *I*nsensitive. Take DELLA, an acronym derived from the amino-acid code. These are pale terms, without potency. Small wonder that they don't stir the mind. And it was not always this way. What could be more resonant than the idea of the atom? Of positrons and electrons, or of the gravitational force? Is blandness the price we pay for the cleansing of feeling from scientific thought? Perhaps we need symbols that sing as vibrantly as the bougainvillaea on the terrace?

Saturday 5th June

A NEW IDEA

Returning home from Mallorca – writing on the plane – sketching remaining impressions. It's been a wonderful few days, my spirits charged by the change of scene. Best of all, I've had an idea.

Thought at last of a way forward for our research. I feel no hesitancy in writing this. There's nothing tentative about it. I know this new idea is the right one. Represents the way to go.

Each morning we sent the children in amongst the shade of the grove to pick oranges. They vied with each other to squeeze the juice that burst out in a torrent. We shared the juice like a Eucharist, drinking in the sunshine, and fly home part orange tree.

There was a pond at the edge of the grove. Within it, frogs that sang at night. Pond-skaters skidding about on the meniscus. A metaphor for science perhaps: a thin skin that has beneath it depths as unknown to us as the world beneath the meniscus is to the pond-skater.

And then on Thursday we went to a rocky cove and beach near Deià. We'd been there earlier in the week, when the sea had been limpid-calm and inviting. But now it was churning, the waves smashing against the rocks and careering up the beach. I stood for a while in the water, dazzled by the light. By fleeting reflections from the fluxing sea-surface, by the brilliant white of the breaking waves. Feeling their backwards and forwards force around my legs as they surged up the beach and then back again. The salt spray in my nose, my shirt wet-cool and clinging to my back. Exhilarated by the energy of it all. Thrilled by a new experience: the sting of pebbles running back with the undertow and smashing into the backs of my calves.

There was the flash thought that *this* is what science is about: the perception of things previously unknown or unfelt. And as a wave sucked back to the sea, its power almost unbalancing me, stones slamming again against my legs, so the idea came. A simple, shining idea. That I should ask Why? Up to now our research has been about How. About mechanism. About how DELLAs regulate the growth of plants. I saw that it's time to transform the how into why. Why do DELLAs regulate the growth of plants?

Sunday 6th June

Back home. It's warm here too. Muggy. And although summer is only just beginning, there's already a sense of things going over. Of things disintegrating, decaying, falling apart. The forget-me-nots already lank. Drying stems with the last blue flowers at their tips, rows of buff and brittle seed pods below.

Tuesday 8th June

Today it is summer without any doubt. It's very warm. Not as warm, or as stridently bright, as it was in Mallorca, but warm none the less. As I write, the tiny dot of Venus traverses the sun's disk, but with no diminishment of light or heat. I'm at once excited and comforted by the sight of the garden from the window. The hazels, oaks, and limes are full now. There's a feeling of completion, of leaf expansion done or all but done. Spires of foxgloves reach for the sky. The sense of present comfort heightened by a sudden shadow of memory of winter's darkness.

Wednesday 9th June

ON THE FUNCTION OF *AGAMOUS*

Light streamed into the room when I parted the curtains this morning. Soon after, I was cycling along those familiar paths to see the thale-cress plant and its graveyard landscape. At last able to see what's happened whilst I was away. Pedalling along in the sunshine. I never find it dull. I do the same journey every time, from the house, to Bracondale, along Whitlingham Lane, then out to Surlingham, several times a month. But I'm never bored with it. It has ritual quality. There is reassurance in the repetition. The familiarity of the scene allowing me to see reality in a new way every time.

I stopped for a while in Whitlingham lane. Picked a buttercup. Traced my fingertip over the surface of each organ of the flower. From the mat, hairy base of the sepals round to their tips. Then on to the silk yellow of the petals, then the pricks of the stamens, and finally the roughness of the stigma. It reminded me of something in Gerard's *Herball*, how 'it chanced, that walking in the field next to the Theatre by London, in the company of a worshipful Merchant named Mr *Nicholas Lete*, I found one of this kind [buttercup] with double flowers, which before that time I had not seen.'

It was doubtless a variant. A mutant form having two layers of petals rather than one. These things are common in nature. Indeed, there are double-flowered mutant forms of thale-cress. They carry mutations affecting a gene known as *AGAMOUS*. Like all genes, *AGAMOUS* is a small segment of DNA, a few thousand base pairs in length. Changes of just one out of those few thousand into another (a substitution of an A-T for a G-C for example) can cause such a mutation. A change that alters the code in the open reading frame, resulting in a premature stop-codon. A codon that codes for nothing. The result, a prematurely truncated chain of amino acids, an incomplete protein. An incomplete protein that doesn't work.

In normal plants, the AGAMOUS protein helps to determine the structure of the flower. Because the protein in the mutant doesn't work, its flower is double. Of course it might have been more severe. A grotesque, monstrous thing bearing no resemblance to a flower. But order is preserved. The mutant simply exchanges one organ for another. A petal for a stamen. It is a subsitution, a change in identity.

The alteration of that single base pair causes the replacement of one precisely defined structure, a stamen (filamentous and transparent with an anther on the top), with a totally different, precisely defined structure, a petal (flat, planar, opaque). And there's a simple explanation for this: the AGAMOUS protein is a transcription factor. It works to control the activity of genes, turning some on, perhaps turning others off, creating a spectrum of gene activity that is a reflection and consequence of its own activity. That particular spectrum contributes to the formation of a stamen. Without it, the stamen cannot form, and a petal is formed instead.

Afterwards, I continued on to Surlingham. To the horse chestnut-shaded graveyard. Cool and peaceful. And in that shaded haven, the damaged plant. I peered in at it through the chicken-wire dome. So much has happened! The thin stem several inches longer than it was. Around it a spiral of flowers, with more buds crowned together at the top. Leaves growing out of the stem, below the lowest flower. I drew a sketch. The stem is held out at an angle from the browning remains of what was once the vegetative rosette. Already bent by its own weight – it may soon perhaps be bending to the ground. The lowest flower is the oldest, the one I first saw open just before going to Mallorca. Whilst I was away, its anthers will have opened and, by touching the stigma surface, transferred pollen, achieving self-fertilisation. Now the anthers are brown and beginning to shrivel. The pod beginning to lengthen. The next flower is also past pollen-shed, but the two above have anthers that are grainy with pollen. And the last flower is less mature again. It's a gradient of developmental age: oldest flower at the bottom, youngest at the top.

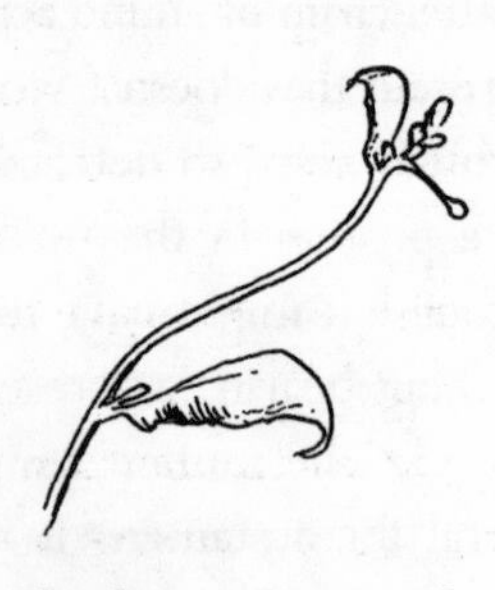

The flowering stem of the thale-cress plant.

So pollination, fertilisation, has begun, and that is of course very good. But I'm disappointed that I didn't see it happen in that first flower, that I was away when it did. And I'm still not convinced the plant will make it. Is it strong enough? The one good remaining rosette leaf is dying. The stem is thin, so wispy that I'm amazed it still stands. And I had expected more flowers to have been produced whilst I was away, more growth, given the prevailing warmth of the air. In particular, with one dying leaf, will the plant be able to feed the seeds that are now forming in these pollinated flowers?

Thursday 10th June

To work. Humid and warm. The sky amazing. A high layer of grey cloud covering the entire expanse of it. The underside not flat like it so often is, but undulating in rounded hills and troughs of similar size that give a strong sense of perspective as their appearance shrinks with distance. An awareness of space that comes from the wide Norfolk sky.

But there is another thing about Norfolk. It rains less here than it does elsewhere in England. There's been virtually no rain here for the last month or so. Yesterday I noticed that the soil the thale-cress plant is growing in was hard, grey, and dusty. Surely the plant, fragile as it is, needs more water? But I've known years in Norfolk where it hardly rained at all from early June until the end of September.

And I'm still occasionally frustrated with the damage that was done. Irritated by images of what could have been. The plant was doing so very well before it got eaten. By now it could have been luxuriant in its growth, with a strong main stem, ten or more subsidiary stems growing out from the buds in the axils of the rosette leaves, all stems branched and covered with flowers that were robust and plump with health. Instead of which I have a plant that is sickly, struggling to make just a few puny flowers.

Saturday 12th June

I think that part of what I'm trying to capture here is the way thought jumps around from thing to thing in the usual course of events, whilst science is often portrayed as having an unswerving linear logic. In fact I, as a scientist, thinking day-to-day about the growth of the thale-cress plant, the experiments in the lab, the DELLAs, etc., jump with ideas and concepts all the time. Last night, for example, I was lying on the bed, looking out of the window at the leaves on the oak tree in the garden, at how they were being buffeted and pushed around by a strong breeze, moving against the static backdrop of the grey, cloud-covered sky of declining dusk light, and not thinking of anything in particular. When suddenly there came into my mind an image of the DELLAs, located as they are in the nucleus of each of the cells that comprise the leaves that were bending and dancing in the eddy and flow of the wind, so that the DELLAs themselves were what was being blown about, moved in space.

And today the cool breeze continues and brings with it heavy showers of rain. Rain is falling fast in parcels shaped and moulded by the wind. At last, some water for the thale-cress plant.

Monday 14th June

A great day yesterday. Wonderful weather, profound music, a visit to Wheatfen.

There was lovely yellow sunshine throughout the whole day. Full but less hard than in Mallorca. With the softening influence of humidity. Clouds white and fleecy.

We went to a lunchtime concert. The Janáček String Quartet. Fantastic playing. Smetena and Janáček. Moving music, expressive of tension. Between the uniqueness of the moment and the development of a durable structure. The struggle for an integrity that matches the potency of a second's vision. Quite lovely. Fragments of sweet lullaby, incandescent shards, insect-like ostinati, snatches of folk tune, all knitted and woven into a robust thing that has relentless momentum towards its end.

Then with the children to see the swallowtails in Wheatfen. Such a sense of momentum, of drive here too. Life driven by water and warmth. So much growth since I was last here. (I wish I could come more often! The last weeks so madly busy at work, there was simply no time.) This year's new reeds already at the height of my shoulder. The beautiful, soft grey-green of them a surrounding expanse. The place hums with insects, scratches with the breeze.

Studded amongst the reeds are glowing yellow clumps of flag-iris. And flying fast but fluttering between the iris flowers are the swallowtails. They fly in bowed trajectories, wobble from one to the next, just above the reeds. A little too fast to allow me to capture a robust image – so that sight is constructed by compiling snapshots of moments of flicker: a split second's seeing of a yellow/black-chequered wing in full sun, a glimpse of a dark underwing, all the time trying to focus on something that has already gone. Much of what is seen is the chance convergence of eye being focused on a particular point and the arrival at that same point of a wing or charcoal abdomen.

On the leaves of the milk-parsley that grows amongst the reeds of the fen there are tiny mounds of swallowtail eggs. And small, crawling, feeding black caterpillars. There's the sense that the reeds, the milk-parsley, the insects, the irises are all part of the same thing, a thing the DELLAs are part of too: the fen.

Tuesday 15th June

ON THE FORMATION OF FLOWERS

A beautiful morning but cooling – the wind moving into the north. A clear blue sky with the merest hint of haze. Leaves shaking and oscillating. The light bouncing off them, flickering.

A few days ago I wrote about how the loss of a single transcription factor can transform the internal whorls of a flower. Turn stamens into petals. Now I'll write about this again, and in particular about how the different identities of the organs of a flower are determined by the activities of particular combinations of transcription factors. This is how the flowers of the thale-cress in St Mary's are constructed. And all other flowers, all over the world.

The way that these transcription factors work is described by a model known as the ABC model. The model was built by looking at variant flowers in which organ identity is changed. In addition to the variant I described previously, there are others in which organ identity is changed in a different way. For instance, there are variant flowers that, instead of sepal, petal, stamen, carpel (the normal), have sepal, sepal, carpel, carpel. As before, this variation is caused by a mutation in a single gene, and once again, this gene encodes a transcription factor. But the transcription factor encoded by this new gene is different from the one affected in the previous variant. By looking, looking deeply at these and other variant flowers, in a way in which looking and thinking become as one, an understanding grew.

So here it is: the understanding, the ABC model. There are three types of gene, called A, B, C. Each type, A, B, and C, encodes a transcription factor, a protein that affects the activities of other genes. Each can be either 'on' or 'off' in any particular developing organ. It is the combination of transcription factors, and the particular texture of gene activity that results from that combination, that gives each organ its identity, that gives it for instance the quality of being a sepal or a petal.

Whilst a normal flower has, in order going into the flower, sepals, petals, stamens, carpels, mutant flowers lacking A function have carpel, stamen, stamen, carpel. Mutant flowers lacking B function have sepal, sepal, carpel, carpel. Mutant flowers lacking C function have sepal, petal, petal, sepal.

Of course the people who first imagined the ABC model had only the strange forms of these flowers to work on. They didn't

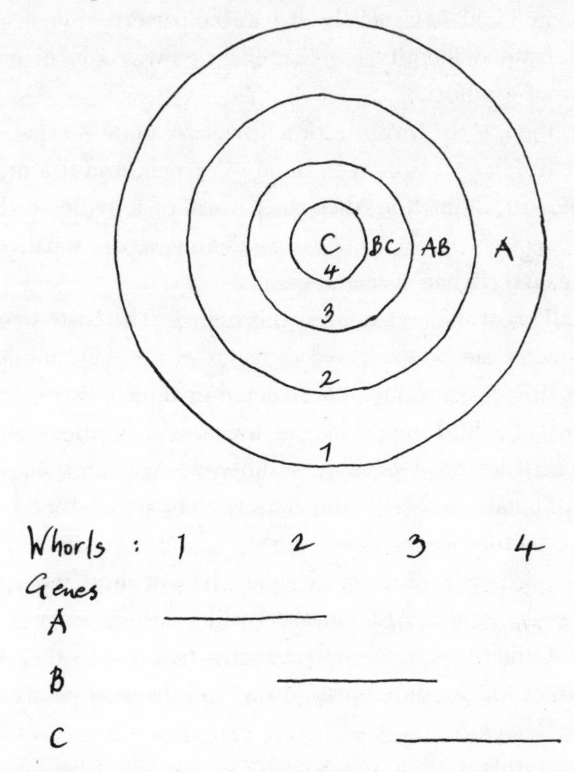

The ABC model. The four concentric whorls of the flower, with the A, B and C gene activities that are characteristic of them. Below, the way in which A, B and C gene activities overlap in adjacent whorls.

then have the framework of the ABC model on which to hang what they saw. So they created the ABC model as a way of making sense of the different flower forms. They made a pattern. First, they proposed that the four concentric whorls of the flower could be envisaged as four distinct territories, 1–4. Next, that the expression of A, B, and C function genes overlapped, as shown in the sketch. With A function expressed in whorls 1 and 2, B in 2 and 3, C in 3 and 4. Crucially, it was the pattern of gene activity that determined identity, so that identity was not an inherent property of a whorl.

Here, then, is the formula for a flower: A alone = sepal, A + B = petal, B + C = stamen, C alone = carpel. And if a mutation abolishes, say, B function, then the pattern of activities within the whorls becomes A, A, C, C – predicting sepal, sepal, carpel, carpel, exactly what is seen.

Like all great understandings, this unifies. The basic structures of all flowers can be explained in terms of the ABC model. Yet different flowers, although constructed in similar ways, are very different. The buttercups in the meadow and the thale-cress flowers on the grave have very different sizes and shapes and colours of petal. So there is much more to be understood before a complete picture of the development of a flower can be said to have been achieved. But we can guess how it must work. Some things, such as the ABC model, will be fundamental to all flowers. Other things, or differential activities of other things, will account for the differences in size and shape of petals seen in different flowers.

To recapitulate, then, the making of the flower, that first thale-cress flower now shaking on its flimsy stem in the breeze. First the inflorescence meristem made the floral meristem that was to become the flower. One in a spiral of floral meristems. An outgrowth on the flank of the inflorescence meristem that then became a dome-shaped meristem in its own right. Then the ABC genes were expressed in the floral meristem, making the ABC transcription factors. These genes were expressed in overlapping concentric rings, with A on the outside and C in the middle, rings around the dome of the meristem like the stripes on a footballer's jersey. These overlapping rings of expression defined four rings of activity: A, AB, BC, C. Within these rings of activity, cells began to form outgrowths on the flanks of the floral meristem that were destined to become sepals, petals, stamens, and carpels. This is how this flower was made.

The understanding shows that it's the activity of a single gene that separates a petal from a sepal. That the one is essentially the other, except for the transforming activity of that single gene. So could it be that petal and sepal are both modified forms of something else? Often in life one thing comes from another. And it's the case here as well. Because mutant plants that lack all the ABC genes make remarkable flowers. Perfect flowers, shaped in a series of concentric whorls. But in the place of the expected different organs, the whorls of sepals, petals, stamens, and carpels, these flowers have concentric whorls of leaves. Sepals, petals, stamens, carpels, all replaced by leaves. These flowers tell us something absolutely profound. They tell us that the ABC genes direct the development of the floral organs away from a basic, a ground, state. Sepals, petals, stamens, and carpels are all modified leaves. Alternative metamorphoses from a common start point.

So where does it all end? If sepals, petals, stamens, and carpels are modified leaves, what are leaves? Are leaves modified stems? Does it go on and on like this so that all living things and all the parts of them are but modified forms of other, progenitive, things, each modified from the next by the simple switch of a gene?

The ABC model is wonderful. A remarkable insight. A product of substantial investment, the result of struggle. Although individuals made the crucial creative inputs, we have collectively, as a culture or society, contributed to the effort to get to that certain place – to see that vision. The tragedy is that we don't collectively own the vision. The model isn't generally assimilated. It's boxed off, separate, something 'scientific'.

How to solve this? I wish I knew. And I don't think that 'public understanding of science' initiatives are the answer. Such things tend to speak in a simplified form of the same language that science already uses. Is it more of a problem of seeing, of vision? Of making images that are resonant, maleable, rich in meaning? Images that sing. How to balance this against the need for neutrality, objectivity, the separation of observer and observed?

The ABC model is achieved. We should now be be retelling it, shaping it to make it sing. By deepening, enlivening, thickening the imagery. Broadening it out from the box within which it was first conceived. Giving it the power of myth. Fleshing out the flatness of A, B, C.

Wednesday 16th June

ON THE GROWTH OF FLORAL ORGANS

And what about our own stuff, the DELLAs, 'relief of restraint', and all that? Anything to do with the development of flowers? Of course. DELLAs are crucial to the growth of the whole plant, flowers included. Here is why we know this to be the case.

I've described the gibberellin-deficient thale-cress mutant before. The flowers of this mutant are very different from those of normal plants. Although they have relatively normal sepal and carpel growth, the elongation of stamens and petals is retarded in gibberellin-deficient mutants. A close look at these flowers reveals short petals and stubby stamens, as if they'd started to grow and then stopped. Presumably gibberellin is needed to complete the progress of growth of these organs.

So the question becomes: if gibberellin is needed for the growth of stamens and petals, does it work via DELLAs? Removing GAI and RGA suppresses the dwarf-stem aspect of gibberellin-deficient plants. I wrote about that previously, when describing the test of the 'relief of restraint' model. But removing GAI and RGA doesn't restore to normal the retarded stamen and petal elongation that are characteristic of gibberellin-deficient flowers. So gibberellin doesn't promote stamen and petal growth by opposing the function of GAI and RGA only. Could it be that the other DELLAs – RGL1, RGL2, or RGL3 – did the job?

Recently, we have answered this question. And it's great to see 'relief of restraint' strengthened and extended by means of experiments. These experiments involved further steps in the painstaking genetic construction of plants that are gibberellin-deficient and that also lack DELLAs. These plants lack gibberellin, lack GAI and RGA, and also lack RGL1 and RGL2. They grow tall and strong (as expected, because they lack GAI and RGA). The most important question was what their flowers would look like. Would they look like those of normal plants, or of gibberellin-deficient plants? When the first flower opened, the result was absolutely clear. The petals and stamens were long, not short. In fact, longer than those of normal plants.

So we could conclude that, together with GAI and RGA, the DELLAs RGL1 and RGL2 play a significant role in the restraint of petal and stamen growth. That stamens and petals grow because gibberellin overcomes the effect of GAI, RGA, RGL1, and RGL2. Most excitingly, the observation suggests that the relief of DELLA-mediated growth restraint is, in a general rather than a particular sense, the way in which the growth-promoting hormone gibberellin controls the growth of plants.

So I can say now how the thale-cress flowers growing in St Mary's have come to be as they are. By invoking first the one model and then the other. First, the identities of the organs of the flower are determined by the mechanism as outlined in the ABC model. Second, the growth of these organs proceeds as outlined in the 'relief of restraint' model. These models represent a real enhancement of vision. Of course I'm proud of the part played by my research group in the achievement of the latter one. But there's more to it than that. I think this enhanced vision brings us, by a tiny increment, closer to the world, to being more a part of it. That's the real importance of it.

To return to the new research question. The why question that hit me on the beach in Mallorca. It's actually a difficult question to deal with. Why questions are always harder than how questions. But I think I'm formulating a way of dealing with it. The beginnings of a way to address it experimentally. Why did plants evolve DELLAs? Presumably because DELLAs confer benefit. So what could this benefit be? Since DELLAs control the rate of growth, the question now becomes: what is the benefit of being able to control the rate of growth?

Friday 18th June

A FLOWER POLLINATES ITSELF

Showers and hot sun. The road steamed (a snatched bicycle ride to see the thale-cress plant again). The world seemingly dominated by yellow: from buttercups, dandelion and coltsfoot flowers. Going along the stony track from Whitlingham to Woods End, I knew it was so warm that one or other of the flowers on the plant would be shedding pollen.

To the shady, sheltered churchyard. The hush of wind in the chestnut trees. I rolled back the dome of wire, knelt, looked at the flowers through a magnifying glass. And there it was. In flower number four. The stamens now at their full length, the anthers at the tops of the stamen filaments dusty with released pollen. Anthers contacting stigma with a gentle touch. The stigmatic surface grainy with the yellow pollen caked amongst its mass of moist hairs.

On the surface of each pollen grain, invisible to the naked eye, is a coat of fantastic architecture. And inside, there are three nuclei. When a pollen grain lands on the surface of the stigma, it germinates. A tube burrows from it into the stigmatic surface. The three nuclei, each containing DNA, travel down the tube. First the vegetative nucleus, responsible for making the tube, then the two sperm nuclei. Incidentally, the growth of that tube is controlled via relief of DELLA restraint. I was thinking about how, at the very moment I was looking, and although invisible to my eyes, these tubes were pushing their way into the stigma and into the tissue beneath. Worming towards the unfertilised eggs that lie within the carpels. How the final outcome of this growth would be the fertilisation of an egg by a sperm nucleus. The beginning of a new plant.

Now I'm back at home, reflecting on what I've seen. This turn in the seasons' cycle. The sense that the whole world is flowering through that thale-cress plant. That there's no concrete division that makes the plant one thing and the earth another. That we only express it that way because that's how we've always seen it. Another way of expressing it is to say that the world itself is flowering.

Everything led to this moment of fertilisation. It is at once culmination, destination, stage, and station in the life-cycle. Yet as I write this I'm aware of a jarring note – the cycle image is so accurate a representation. There's no defined point at which it stops or starts. No real climax. No most important bit. So perhaps I shouldn't have been so partial in my excitement with the anticipation and moment of fertilisation. My pleasure, my sense of reverence, is better spread without favour through the cycle as a whole.

Saturday 19th June

ELABORATING THE NEW IDEA

It's much cooler today. A breeze from the north-west. Fragile bursts of sunshine, dark showers, alternate. The wind and pulsing light bring a sense of instability. An instability that's inherent in the beauty.

We're approaching the middle of the year. And I'm going to make a prediction about the second half of it. That these pages will become increasingly preoccupied with the new idea that has been growing in my mind. The idea is that DELLAs connect the plant with the world outside. That, I think, is the benefit that DELLAs confer. They enable plants to grow at rates that are somehow appropriate to prevailing conditions. They allow the world to control the growth of plants by increasing or reducing restraint. The question is how to test this. I think that much of the remainder of this year (and doubtless more time beyond it) will be taken up with the design and completion of appropriate experiments, the writing of the papers that describe the resulting observations. And I'm so excited that at last I can see a way forward. A route to a further expansion of vision.

Sunday 20th June

ON CONNECTIVITY

Out again to St Mary's. This time I'm so sharply aware of the thale-cress's fragility. A single thin stem, held at an angle, bent by the weight of flowers.

The recent growth has been slow. Despite all the showers, the soil remains grey and dusty. The plant looks parched. But what with the paucity of water in the earth, and such a scrawny stem, it's a miracle the plant is growing at all. I think it is only just making it through.

Cycling home, I catch sight of a purple buddleia in a garden. The first I've seen this year. And suddenly I'm on a train to London. An intense memory. A humid afternoon several summers ago. An urban landscape. A desolate industrial hinterland through which the train passed on its way towards the centre of the city, an arrow to a heart. It was then that I first became aware of the great stands of buddleia that were incongruously in bloom in the shingle beside the railway tracks, in the most infertile looking of places. The train passed plant after plant as we travelled into London. It seemed that the harsher the landscape, the more vigorous the buddleia became, its roots presumably penetrating with ease the flaking bricks and mortar, the crumbling concrete.

I found the plants so striking, was so affected by the cumulative impact of the passing radiant blooms, that the buddleia began to invade my thoughts. Soon I saw a living London, great purple arteries radiating out from its heart, miles of railway tracks lined with flowering buddleias.

Shortly before our arrival at Liverpool Street Station the train slowed, then came to a halt, the silence broken only by the ticking of the stifling heating system of the carriage. Ahead rose the walls of a dark tunnel which the train was about to enter. The only living thing to be seen was a buddleia plant, just inches away, rooted in the mortar at the base of a wall of black bricks.

I looked through the thick, dusty glass of the window at the plant. I could see its tall, bowed, stretching stems, each topped with a conical spire of purple flowers, small and soft. Doubtless they smelled of honey, of sweetness from a bitter soil. Other flowers were beginning to go over, turning brown and shrivelled.

The stems had tiered leaves, pitted and spotted by the city's grime, splaying out on either side at regular intervals.

As I looked I reflected on how these long cylindrical structures, these stems swaying in the warm wind, the tops of their flower spires describing circles in the air, were the shape of their function. That they were a collection of tubes, bundled together, and made of cells especially adapted to the business of being conduits, placed end-to-end like the segments of a drain.

As I sat in that stationary railway carriage I remembered a time at school when we'd studied transpiration, the mechanism that sucks water into the leaves and stem of the plant. We cut a branch from a tree, a sycamore I think, and immersed the cut end in a bucket of water. Then we took the bucket into the lab and connected a short section of rubber tubing, itself full of water, to the branch. At the other end of the rubber tubing was a glass tube, also full of water. We briefly raised the free end of the glass tube above the surface of the water to introduce one tiny bubble of air. And then we watched, marvelling at the speed with which the bubble, within the space of a few minutes, travelled along the length of the glass tube and disappeared into the rubber.

It was the connectivity of the thing that most amazed, the connectivity of the different parts of the plant. How water was evaporating from the cells of the leaves, passing out of the leaf through the epidermis and the pores, and how the effect of this on the vessels of the stem was like suction on a straw, so that the movement of the bubble we could see was the flow of the stream of water being sucked through them by the leaves.

A sudden flash of colour, a flicker of red and white, pulled my mind back to the landscape beyond the window of the railway carriage. A red-admiral butterfly alighted on one of the buddleia spires and then fluttered from flower to flower. Finally, it settled, and extended the long, coiled tube of its tongue into the depths of the flower, searching for nectar. It seemed to me as if an electrical circuit had been completed. Tubes were connecting the butterfly to the earth and to the sun, and it was feeding on them. As if we are all part of one huge organism connected by a network of tubes and conduits to the earth and to the sun.

Wednesday 23rd June

Today I drove out to Holt to see the dentist. In wild weather. By the side of the road a birch tree: leaves flickering violently, branches bent parallel to the direction of force of the wind. But between deep green roadside hedges, damp and tall on either side of the road, there was shelter. Today has an autumnal feel. Are we already propelled in that direction, so soon after the solstice?

The last few days I've had in my mind an image that life has two surface layers to it, the main body of existence sandwiched between the two. A bottom layer that is dark, troubling, full of things to fear; a top layer that glitters, is irridescent, beautiful, a source of joy. And then there's all the rest in between. Both surfaces are always there: shit and honey. You can't have the one without the other.

Thursday 24th June

In the evening (cool, calm after showers), managed to snatch a visit to St Mary's churchyard. Dark in the shade of the trees.

The first flower of the plant is finished. Petals, brown slivers that hang limply; stamens/anthers shrivelled; but at the centre the carpels are extending, pushing outwards to form the fruit. Inside the fruit, the seeds are forming. How many? And inside each developing seed, a tiny embryo. The next generation.

Friday 25th June

ON THE FORMATION OF EMBRYOS

The formation of the embryo is miraculous. That one single cell can build a multicellular organism out of itself.

This first cell, termed the zygote, is formed by the fusion of sperm and egg nuclei in the process of fertilisation. Controlled proliferation from there on, during the course of ten days or so, forms the embryo. At the very beginning: the establishment of polarity. The embryo is polarised. It has a top and a bottom. At the top: the shoot meristem. At the bottom: the root meristem. Populations of cells that will make the shoot and root of a plant.

So how to get from one (single-celled zygote) to the other (multicellular embryo)? By a series of cell divisions and expansions. To begin with, a division of the one cell into two. It's at that very earliest moment that up and down are established. The bottom cell is destined to become a structure that attaches the developing embryo to the inside of the growing seed. The top cell

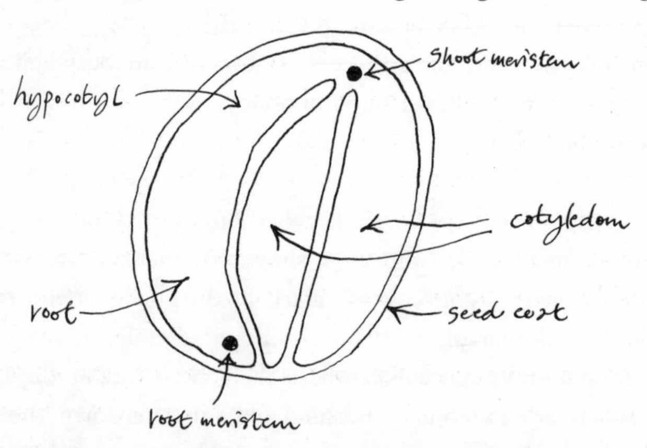

Diagram of a thale-cress seed. Folded inside the seed coat is the embryo. The positions of shoot and root meristems are indicated with black dots.

and its daughter cells will form the embryo. A series of cell divisions and expansions, coupled with the beginnings of the adoption of individual cell identities, elaborates the embryo from that founding top-most cell.

So here is an abstracted image of the mature embryo. At the top, two cotyledons (embryonic leaves in which most of the seed food reserves are stored), on either side of the shoot meristem. Beneath the shoot meristem, the hypocotyl (embryonic stem containing vessels for water and nutrient transport). At the bottom of the hypocotyl, the embryonic root with the root meristem at its tip. That's all. But it's potent. An elemental plant. One end destined to be above ground, the other below it.

Sunday 27th June

There has been a succession of westerly gales and storms. Blown all the way from the Atlantic, yet still with enough rain to drop with huge force on Norfolk. But is it enough?

Yesterday, heard again, for the first time in ages, the *Rite of Spring* (Stravinsky). The awesome velocity of it: from splintered dawn chorus through to the terrifying violence of the last dance. The brutal in Nature. It made me think of what I'm trying to capture here – I wonder if I've missed an aspect? I've caught awesome/beautiful okay, I think. But have I really caught awesome/terrible? Now that we're past the solstice there's a downhill plunge in sight: death for the plant, destruction, winter darkness.

Since we're just past the solstice, the halfway point, perhaps it's time to take stock. To review. I've described the growth of plants, the scientific understanding of that growth, how the work of my research group has contributed to that understanding. And I've set it all against a backdrop of the seasons, of shifting climates, the instabilities and sudden flashes of consciousness, the diverting jerks of memory or attention-grasps of immediate perceptions. It seems good. It begins to present the science as part of a whole. Yet I'm aware of dissatisfaction. A feeling that there is further to go. The word *backdrop* jars. Perhaps still greater integration can be achieved?

Tuesday 29th June – Kiel, Germany

A brief visit to Kiel – an annual meeting of a network of linked EU labs that I am part of. Here via plane from Stansted to Lübeck. Then train though northern Germany from Lübeck to Kiel. The sky light grey, yet light. It's cool, damp, with sporadic showers. The landscape sparsely populated. Travelled through dense, shady forests past slate lakes, wide, rolling fields of wheat and barley. Seeing the barley, its beauty, its collective movement like a water surface in the breeze, has put a thought into my mind. That I should revive our recent work on barley. Continue the investigations of the barley DELLA. After all, the circumstances are fortunate. A group in Scotland has recently begun an effort to identify new mutants in barley. When I return to the UK I'll contact them. I love the idea of working with this beautiful plant again, in parallel with thale-cress. We were working with it a year or so ago, but the project faded because I wasn't sure how to make further progress with it, couldn't see the next step. Now I think that perhaps I can.

JULY

Thursday 1st July

THE weather in parcels and bursts. Sunshine, then dark cloud-shade; light, then vertical rain in fat drops. The change of scene directed by the westerly progress of the clouds. It's cool and humid.

Although I was previously concerned by the dryness of everything, and although we haven't had any sustained rain recently, still this continuing spell of sunshine/showers (as though it was set solidified in its volatility, it began Monday and has lasted since) is providing moisture enough for the continued growth of the thale-cress plant.

Friday 2nd July

This summer is rushing by. I've missed so many things. Cycling past them on my way to work. No time to really look at them, to stop and see. Sometimes life is like that, a series of rushed glances. Bouncing from one thing to another.

But at least I did get to see the thale-cress plant this evening. Felt the peace of the darkening churchyard. And there's no question but that the plant is slowing its growth, is close to stopping. The stem, that one thin stem it managed to make, is I am sure little longer than it was the last time I saw it. At its tip is an aborted flower bud – messed up and wizened. This is significant. It means that that stem-tip inflorescence meristem has finally become a floral meristem. Since a floral meristem is determinate, has an end, will make a flower and then stop, this means that the stem will stop growing.

So that's it then. The growing phase is finished. The plant's entire future is carried within the scant dozen seed pods that fragile stem has managed to make. The sketch records the point at which it has stopped. To think how many pods there might have been. Could have been several hundred.

The pods on the flowering stem of the thale-cress plant.

The first five pods are full enough, they probably contain about thirty seeds each. The rest, well, they are pretty short. About ten seeds apiece, I guess, or less. So roughly 200 seeds in all. When a normal thale-cress plant, grown in good conditions, would make upwards of 15,000.

That rabbit, or whatever it was, changed the world when it all but demolished the thale-cress plant. Had its own impact on the future of the thale-cress species: reducing the representation in future generations of whatever particular forms of genes this individual plant contains.

But at least there *are* seeds in those first two pods. The pod walls have the slightly rounded indentations that sometimes occur when the seeds are filling out and the pod wall tucks and wraps around them. And the seeds in these older pods must now be close to maturity. In each the embryo will have developed and expanded through its various phases until it has filled the space of the seed. Each entire seed is now a tiny progagule. A plant in unit form: top meristem; bottom meristem; vessels connecting the two; nutrients stored in cotyledons for future seedling establishment.

Another stage in the cycle is coming to a close. The plant is at the beginning of the end of its life. But what new is there to be said about that? That there are only a few fleeting moments between birth and death. All continuity vested in the succeeding generation. Part of an unbroken chain, yet individual existence a transient flicker. We all know this. We can cry, sing, or dance about it. Share the beauty and the terror of our lives. Or suppress the thought of it.

Sunday 4th July

ON SEEING AND PERCEPTION

Yesterday, I played in a concert with the children, them on their violins, me on the piano. I was suddenly struck by the nature of people's individual sensitivity to music, and a possible parallel with the subtlety of grasp of scientific things. Both are all about nuance, I think, about shadings that I suppose some are aware of but others not.

I imagine that the fact that I have musical awareness is partly a product of experience. I began learning the piano at the age of seven, and have been playing more or less continuously ever since. Forty years of ear and mind being attuned to the attack and decay of sound, to the precise discrimination of its qualities (of, say, the difference between a staccato that is a stab or a droplet), to the use of the hard and the soft, the mixing of bitter and sweet, and to the slightest stretches or diminishments of time. And to responding rapidly – to shaping my own playing to make a consonant whole of the sound made. It is I think generally agreed that to play music demands a special facility that is partly learned but also dependent on the personal, the idiosyncratic.

Science is like this too. I see the DELLAs in a particular way that is unique to me. Of course other scientists have their own intricate pictures of those same proteins. But these pictures will be different in detail, in emphasis, from my picture. Most people, because they haven't been steeped in it all, cannot have this depth. Probably for them a protein is a flat or a smooth thing that has little in the way of texture or underlying detail.

Nor does all this cover only the invisible or the conceptual. The observational is part of it too. I am aware of the growth and development of the thale-cress plant, of tiny changes in it, and I know that particular awareness helps me to see.

Today, we took Alice and Jack to Stiffkey, on the north Norfolk coast. To catch crabs. Here there's a long track that extends out towards the sea across the salt-marsh. A wooden bridge over a creek of brown water. As the children began to tie pieces of bacon to their lines, I walked on into the vast flatness. Leaving the path, I stepped on to a mat sprung with samphire and sea-lavender. The scent of sea salt. The world seemed as if a half-sphere: flat land extending into the distance, dome of sky above. The harsh babbling and scratching of skylarks high in the air: an expanse of song that seemed to reflect the horizon's extent. And suddenly, I could see the next experiment in my mind.

A few days ago I wrote that I wanted to address the question of why plants have evolved DELLA-restraint, of what benefit it brings. But that I didn't know how to address this question experimentally. Here, in the salt-marsh, it was clear in an instant. The salt-marsh is a place of extremity. Regularly the sea floods the ground, loading earth and mud with salt. Although the plants that are adapted to life here thrive in such salinity, other plants do not. Thale-cress, I am sure, would not grow well in such adverse conditions. And that is the experiment. In essence, to recreate a salt-marsh in the lab. Then use it to test if DELLAs somehow regulate the growth of plants in response to adversity. If they do, then the evolution of DELLA-restraint is explained. It is so obviously the thing to do that I cannot for a moment understand why I couldn't see it before.

I returned to find Alice and Jack excited, each with buckets full of crabs of varying size, dull red and brown. The water in the buckets churning with their movement. After a while we released them. Watched them sidle in unison down the bank, plop into the dark water of the creek, seagulls wheeling and screeching overhead.

At the same time as beginning the search for *why*, I think I also need to discuss the how in a little more depth. Previously I wrote of the 'relief of restraint' hypothesis, that DELLAs restrain the growth and proliferation of plant cells, that gibberellin works by overcoming that restraint. And I provided genetic evidence that plants lacking DELLAs grow, even if they lack gibberellin. What was still missing was any real understanding of how gibberellin overcomes the growth-restraining effect of the DELLAs.

The first suggestion of how the how works came from experiments done in another lab. In these experiments a gene encoding a 'fusion protein' was expressed in the cells of thale-cress plants. This fusion protein is two proteins joined together. It consists of a DELLA (RGA) with another protein, GFP (green fluorescent protein), fused to one end of it, making GFP-DELLA. GFP does exactly what it says: it glows green when excited by ultraviolet light. Thus GFP is a marker: is used to detect the location of proteins that are fused with it. Looking at seedlings lit with ultraviolet light and expressing GFP-DELLA through a microscope, the fusion protein glows in the cell nucleus. In itself, this wasn't surprising. It was already known that the DELLA protein sequence has features characteristic of transcription factors. Transcription factors work on genes, and genes are contained in the DNA in the nucleus. So it was expected that GFP-DELLA would be nuclear. But now a very important experiment could be done. The question underlying the experiment: what would happen to the nuclear GFP-DELLA in the cells of plants treated with gibberellin? The result of this experiment was visually arresting and strikingly informative. After a few minutes' exposure to gibberellin, the glow generated by GFP-DELLA disappeared. The nuclei transformed from bright, shining globes to pale disks, dim shadows of their former selves.

This single experiment marked a definitive deepening of understanding. It showed that gibberellin overcomes the growth-restraining effect of DELLAs by causing them to disappear. In retrospect (of course) it seems obvious that it should be this way. That this property is entirely consistent with the predictions of 'relief of restraint'.

So now there was the next question. Another how question. If gibberellin causes the disappearance of DELLAs, how does it do it?

Monday 5th July

SETTING UP THE NEW EXPERIMENT

But to return to why questions. Today we discussed how to set up the salt experiment. It's so very simple to do. We have seeds of two thale-cress lines. The first line is our control: normal plants. The second lacks four (GAI, RGA, RGL1, RGL2) of the five DELLAs. These seeds will be germinated on agar medium containing salt (our essential salt-marsh) or not (control). Then we'll watch, compare the growth of the two sets of seedlings. So exciting that such an easy experiment is so potentially revealing. And rapid too – we should have a result in just a few days.

In the evening I drove out to see the plant in St Mary's. It's clear that it's dying. I was struck by an incongruity. That it's dying amidst the background of vigorous summer's growth. Amongst the dandelions and daisies in the graveyard. I picked a dandelion, smelled its delicate smell, touched the ring of milk-white sap around the break in its tubular stem. And since the thale-cress plant is dying now, I took away the chicken-wire.

A commentary on the current state of the remaining leaves and stem. The rosette leaves are now all brown and shrivelled. The stem-leaf is more yellow than green, especially at its edges. The stem to which that leaf is attached was always frail and now looks wispy-thin. But it would be a mistake to imagine that the body of the plant is simply disintegrating. The process is neither so random nor so bereft of purpose as that. As the leaves die, so they are dismantled. When they were first constructed, the plant invested resources in their building. Now it is reversing this work. Breaking up the cellular constituents of the leaf, reabsorbing them into the stem. The cells of the stem-leaf have begun a phase of self-destruction, destroying and digesting themselves, releasing the materials from which they were built. These liberated materials flow into the phloem vessels, pass through the stem, and enter the seeds growing in the pods. The plant destroys itself to feed its children. The new eating the old. Sacrifice/Eucharist. A familiar refrain.

Wednesday 7th July

MUTANT DELLAS ARE STABLE

Something I forgot to mention in my previous entry about GFP-DELLA. About the difference between GAI and gai. A few days ago I described how GFP-DELLA disappears in response to gibberellin. So there was now a question concerning the behaviour of DELLAs that lack the DELLA region (like gai). How would such a mutant DELLA behave when made visible as a GFP-fusion? The result was

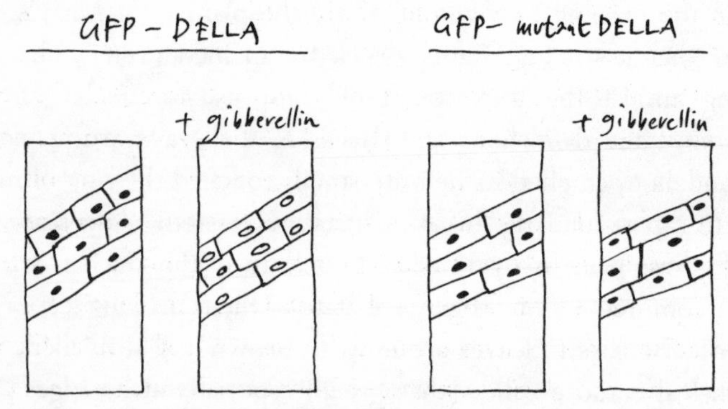

Gibberellin causes disappearance of GFP-DELLA but not GFP-mutant DELLA. The panels show cells of thale-cress roots, with nuclei inside. Filled nuclei are fluorescent due to GFP-DELLA (or GFP-mutant DELLA). Empty nuclei are non-fluorescent because gibberellin causes disappearance of GFP-DELLA.

clear. This protein was stable. As shown in the sketch, it didn't disappear in response to gibberellin. So there's a perfect correlation. Mutant DELLAs that make plants dwarf and resistant to the growth-promoting effects of gibberellin are themselves gibberellin-resistant. Gibberellin doesn't make them disappear.

Thursday 8th July

The last few days there's been lots of rain. It's exciting to think about it: bringing life to the thale-cress plant (at least to the seeds maturing within its pods), to the garden, to the fen. Reminds me of a night driving home in the rain in Ireland from a concert (Messiaen: *Quartet for the End of Time*) in Bantry House. A warm, wet night: moisture in extremity, the air seeming as wet as the rain itself. Suddenly a tiny frog hopped on to the road and into the edge of the beam of the headlights, shone for an instant, then leaped back into the grass of the verge.

Sunday 11th July

ANOTHER IDEA

The ideas are really flowing now. I've just thought of a completely different way of testing if DELLAs regulate the growth of plants in response to the surrounding world. It's to do with gravitropism. The way in which the seedling roots of thale-cress and other plants follow the direction of gravity, grow towards the centre of the earth. Might it be that DELLAs are part of this mechanism that connects plants to the earth, that enables them to grow in the appropriate direction? The test is appealingly simple – we must do it.

And I'm enjoying the summer, despite the cold and the wet. At last it seems that the 'scientist's block' is over. This telling of the tale of the thale-cress plant in St Mary's has helped me to break through that barrier. Although both of the major moments in defining this new path occurred close to the sea, not in St Mary's itself, their occurrence was unquestionably part of a broader context: one in which I'm thinking far more about the world beyond the plants themselves than I previously was. That context is the product of my observations of the plant in St Mary's. Studying one small plant and its place in the world has brought my science out of the confines of the lab, into reality.

Tuesday 13th July

At last I've escaped to Wheatfen. Sitting on the wooden seat beneath the willow. Looking out across the fen. The weather: cool, a soft breeze. The sky mostly covered with cloud, a few blue patches.

Today I'm strongly aware of a sense of relief. The sounds of the warblers in the reeds, chiffchaffs, a distant blackbird, wood pigeons, are calming. A wren plops and flops. Bubbles in the willow above my head. Its wings making a sound that is part buzz, part moth-flutter. The relief grows, expands. Peace.

It's too long since I last sat here. The reed-bed is changed out of all recognition. In the winter the brown dominated the green. Now it's the other way around. The stems of the reeds are shooting up with the regular rhythm of their repeated structure: stem internode, leaf, stem internode, leaf. Countless in number – thousands, millions of individual reed stems all merging in my sight. In places strangled by nettles. It all seems so tangled, so chaotic. The lines of stem and leaf interpenetrating and sticking through one another's space. Not all standing straight, many beaten down by the rain of the last few days.

Closer to the wood the reed-bed is full of flowering meadowsweet. That's a change too – it wasn't out when we came to see the swallowtails. The scent thick and heady. A few tortoiseshells flicker.

I feel fortunate. And suddenly I think: can it really be that DELLAs are part of all this? Even to me there seems to be a disparity. Between the way in which the body responds to the scene: the breeze, scent, buzz of flies, curious scraping chatter of the warblers. And the thought of those imperceptible, unfeelable DELLAs. Yet my head tells me that DELLAs are as much a part of the scene as anything else is. That without them this scene would not be.

Then on to St Mary's. The trees that line the churchyard's edge, in order: lime (seeds small, dangling green globes); then horse chestnut (small, spiky, round fruit cases already – seems only yesterday that these were flowers!); lime; lime; horse chestnut. Dense, brooding canopies of leaves.

In the last day or so, the first of the plant's seed pods has shattered. It's not surpising that it has. It was looking yellowish and ready to go last time I was here. When a pod forms, it's made of two thick coats, known as valves, originally the carpels of the flower. The valves enclose a thin inner membrane known as the septum, to which the seeds are attached in two rows, one on

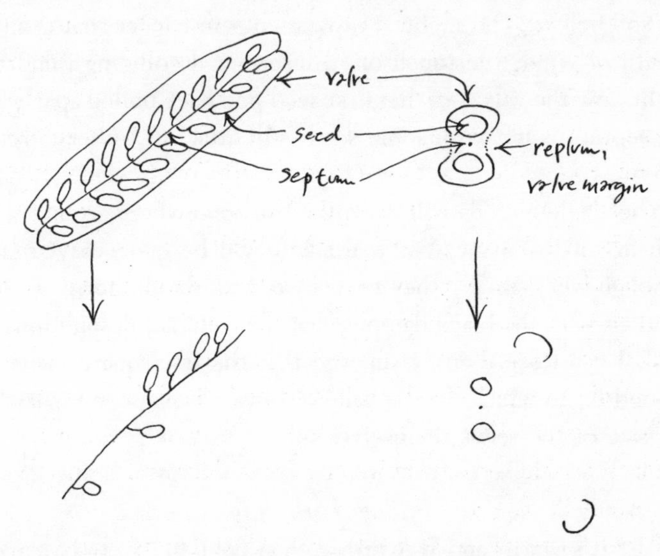

A thale-cress pod in long view (*left*) and cross-section (*right*) before (*above*) and after (*below*) the pod shatters. Shattering occurs when the valve margins fracture and the valves fall away. The exposed seeds then begin to detach from the septum, and fall to the ground.

either side. The valves are linked to one another along the length of the pod by a patch of cells known as the replum. And at the edge that defines the replum, the margin that connects replum and valve, is a thin line of fragile cells known as the valve margin. The entire pod structure forming a case that encloses the seeds.

The identity of the valve margin is defined by particular transcription factors. Indeed there are thale-cress mutants that lack these transcription factors, and that therefore fail to make valve margins. The mutant pods these plants construct don't shatter, because they lack the margins. When a normal pod dries out and reaches maturity, the cells of the margin snap and the pod shatters. The valves are pushed apart by internal pressure. As the pod shatters, the seeds detach from the septum and fall to the soil below. The slightest movement can trigger shattering: a breath of wind, the touch of an insect or a colliding raindrop. Right now the valves of that first seed pod have pulled apart, and the septum is left with some seeds still attached, others already fallen away.

Those fallen seeds will be in the soil somewhere around what remains of the plant. You'd think they'd be quite easy to see: although very small, they're brown and should form a clear contrast with the black dampness of the soil. Yet despite looking hard, I can't see them. I suppose that the few square inches of ground on to which they've fallen is a vast expanse compared to the size of the seeds themselves.

Friday 16th July

At last it's warm and wet rather than cold and wet. Last night there was torrential rain. A roaring sound from the accumulated impacts of massive, speeding drops on the roof, lawn, and trees. This morning the air is humid and soupy-mild.

Getting ready to leave for work, I noticed that the comfrey is dying. It shoots up all over the neglected parts of our garden with great vigour in the spring. Yet now there are patches of blackening rot on stems and branches. The seed cases dry and brown. Already there are the signs of summer going over the edge. Despite the warmth.

Saturday 17th July

ON HOW GIBBERELLIN PROMOTES THE DISAPPEARANCE OF DELLAS

Such exciting weather. The sky is absolutely clear. Totally transparent and without haze. This afternoon I watched a towering blackness of cloud move towards us with slow inevitability. The edge of it so sharp, so clearly defined. A thin line: blue on one side, black on the other. The roll of distant thunder. A change in the space of a few minutes from bright, strong sunshine to the profound shade cast by the cloud. Then the force of the rain slamming into the gravel in the garden, a sudden wild wind in the trees, the lightning flash and cracking roar of the thunder. After ten minutes it was all over, departing with a receding coda of rumbles and thuds.

A further note on the growth of plants. Gibberellin causes the disappearance of DELLAs from the nuclei of plant cells. The way in which gibberellin causes this disappearance is partly understood and partly mysterious. What is certain is that gibberellin somehow marks the DELLAs. It's what chemists call a modification. Some extra bit of molecule, perhaps something as small as an additional grouping of a few atoms, is added to the DELLA (itself a very large protein molecule) in response to the gibberellin.

How the gibberellin causes the attachment of the modifying group to the DELLA remains a mystery. But it is thought that this process marks the DELLA protein for destruction. That once marked, it will be destroyed. On the large scale, the marking of multiple DELLAs results in mass disappearance. The result: growth is promoted.

Monday 19th July

It's working! Today we took a first look at the salt experiment. And the result is clear already. As we had guessed, salt inhibits the growth of normal thale-cress plants. These seedlings are shortened, stunted in their growth. Slower growing than those grown in the absence of salt. But the crucial thing is that the DELLA-deficient mutants don't seem to care. They grow about as fast in the presence of salt as they do in its absence! The idea is tested and has survived the test. DELLAs enable plants to regulate growth rate in response to the environment in which they find themselves. Plants lacking DELLAs cannot do so. This tiny experiment changes everything. Now we have something to build on, can really move forward.

Wednesday 21st July

Travelling again. To Dundee for the rest of the week. Purpose of trip: to screen through a huge field of barley plants looking for new mutants. Apparently there are twenty-five kilometres of walking to be done! Row after row of plants, looking for favourite things: dwarf plants; plants that are too tall, too spindly-thin. Will be interesting to see if my energy fades with the hours of walking and looking. And I know from experience that if the enthusiasm wanes, less is seen, things are missed.

It's a long journey. Engineering works in various places, so: Norwich to Ely, Ely to King's Cross, King's Cross to Dundee.

Norwich to Ely. Through flat fields, gradations of yellow, golden, brown, fields of barley and wheat. Sky full of patchy clouds. It's a little warmer today, rather humid. But still we haven't had any really warm summer weather, and today I'm pleased about this. It makes it likely that the barley will still be green, which will in turn make it easier to spot the dwarfs (they tend to be darker, bluer-green, and so stand out).

I've been thinking a lot about the consequences of Monday's result. Previously, it was thought that plants grown in adverse environments grew poorly because they are 'sick', weakened by bad conditions, their metabolism compromised. The new observation means this picture is incomplete. The inhibited growth, at least in part, is something the plant is doing to itself. It is an active, regulated thing rather than a passive-response thing. And it is DELLAs that enable this active restraint of growth in response to adversity.

So far so good. Not hard to think about. Now comes something harder. Our understanding of how plants perceive adversity is rather messy. At least two distinct systems of hormones/transcription factors are thought to be involved. Somehow these things must relate to DELLAs. Dissecting out these relationships will not be easy, and the end result is unlikely to be a simple linear story. It might be the kind of story that is difficult to write and difficult to read. A challenge.

Ely to King's Cross. Soon I'll be creeping from London to Dundee through a thin skin of atmosphere, an infinity of lifelessness above. What is it that gives this small collection of atoms, the layer that clothes our globe, the quality or property that we call life?

I sometimes wonder if we began again at the beginning of science, let it run again from the start, would we have the same picture we have now? Is there a step-by-step inevitability to the development of that picture? Modern molecular biology, for instance, is rooted in atomic theory. Somehow, this dominates our current view, so that it can seem as if this molecular reality is more important than the larger-scale levels of organisation.

Certainly there is this flavour or bias to what we do. Does this bias limit our vision? If we started again, would the picture be based on a different bias?

King's Cross to Dundee. A short doze, cup of tea, and now I'm having fun. Thinking how to structure more salt-growth experiments. The world whooshes by as I look across fields and towns, see church steeples in Lincolnshire stone, sheep, wheat on the edge between green and yellow, copses, solitary craggy oaks, pylons, and cooling towers. And I'm thinking up connections, ways to structure the flow of ideas, ways to make it all work.

The further north I get, the greener the barley becomes.

Over the Forth Bridge, over the River Tay, into Dundee.

Saturday 24th July

Back on the train, returning from a lovely couple of days hunting for barley mutants. The weather has been kind, bright (but not too bright to see), a mild breeze that felt good on the skin and gently swirled the softly hissing sea of barley heads that surrounded me. The heads shimmer: the awns slightly reflective of the light and, whilst still green, on the verge of browning. The field is on a slope above Invergowrie, and from it the view across the firth of Tay is of the Kingdom of Fife: rolling hills, trees and cattle, patches of cereal fields, all seen across the estuary. Beautiful in the swiftly changing intensities of light, patches moving as if hovering over the landscape.

Hunting for mutants – for short-strawed, blue-green dwarf types; or for taller, spindly, slender types – is such fun. They are very rare: you can walk past thousands of plants and see nothing. But then, suddenly, almost when hope is fading, you see one. The moment of discovery is brilliant. As intense as splitting a featureless slab of shale and exposing a fossil. Something that's never been seen before. Each one having the potential to reveal something new about the growth of plants. In two days' hard work I found fifteen. In the evening I was exhausted and a little euphoric after a pint or two. Mood affects it: be positive and you will find, be despondent and you will see nothing.

Yesterday I looked up and away from the barley for a moment and saw a single tree outlined on the crest of the hill to the west of the field, a green ball silhouetted against the mottled blue and grey of the sky. Looking back at the barley I still saw the tree as a picture in my mind, and came a step further in the understanding of what I'm doing, or trying to do, with this writing. That the writing is an attempt at the merging of minds. That I want to make you, my reader, whoever you may be, see the same pictures I can see. I think there's a growing need for us to share and see the same images.

This new step in understanding of what I'm doing, this advance in awareness, implies a change. A development. Now I'm writing for you, whilst before I thought I was simply writing for myself and for my children. But perhaps I've been veering, tending towards this for some time. And perhaps we can find ways to merge the unshared visions that divide us.

Sunday 25th July

A VISIT TO WHEATFEN

It's warm. Not hot. With a cooling breeze. Shady and humid in the wood. There's a bank of nettles: tall, segmented towers. Straining in the shade and parallel with one another. My foot (bare in a sandal) stung for venturing into them. Then an insect (a mosquito) bites my arm. First the sting, then the bite. I imagine that I'm being penetrated by the wood. Injected.

The stinging hairs of nettles are single long cells, needles with bladders at their bases. The bladder is held within a clump of other smaller cells that rise above the surface of the leaf, and contains a cocktail of irritants. The hair itself is a capillary of the finest bore. Delicate, brittle as glass. The touch of my foot shattered a few of those fragile capillaries along lines of pre-determined weakness. Exposing the sharp edges that then penetrated my skin. With my body pierced, the bladder's contents flowed into me. Molecules that attack, that generate the itch and sting. And then another penetration of my body by the insect that bit my arm, its mouth-tube probing for my blood.

Later in the fen. A reed stands beside me – segmented, leaves sticking out to the side. On the top surface of the leaf are lines of green aphids (with an occasional coral-pink one amongst them). They lie in files parallel with and above the straight lines of the veins in the leaf, tubular mouths penetrating the phloem and sucking nutrients from the plant. It seems to me today that the scene is of a network of penetrating and connecting tubes that link the living creatures in the fen and the wood, knitting the landscape into one.

Continuing on the subject of tubes . . . inside the sheath of the reed is the flowering stem, pushing its way through the cylinder that the leaf sheaths make. I know it's there although I can't see it. Soon it will poke its head out of the end like a chimney sweep's brush, and the whole fen will be topped by a layer of feathery flowers.

Wood pigeons make their soft calls. My toe still itching from the sting, needling thought. A tiny shining brown frog leaps away from my feet into the mud and wet of the reed-bed. I hadn't seen it before it moved and am shocked by the suddenness of it, seeing it first as a leaping of a piece of the earth itself.

Monday 26th July

Yesterday's experiences were so lovely that I had to return today. On the way, I cycled past fields of tawny wheat. Some already harvested, the fields stacked with bales of straw. In another field rows of potatoes in flower: petals white with a hovering purple touch.

Wheatfen car-park is empty. The wind ruffles the trees – birches, conifers, beeches, oaks, the hissing sound of leaves rubbing each other with smooth friction, the rush of air channelled through the spaces between twiglets. The day is mild-humid. Clouds, lumpy white wool, slow-moving, suspended, blue patches between them revealing streaks of flat haze above. Then into the wood. All that remains of the recently vibrant carpet of bluebells is brown. Stems dry, brittle, carrying empty seed pods, lying flat on the ground above a layer of dead oak leaves and disintegrating mould. And I also notice the blackberry flowers. Petals white with purple hint but many flowers gone now, replaced with conglomerations of tiny hard green spheres, a ring of withered stamens below. Bumble-bees pick out the remaining open flowers – elliptical flight with final precise realisation of target.

I walk though the woods and out into the fen. I see common reeds, willow-herb, white convolvulus, meadowsweet, vetch. The old spiky reeds (last years' flowering stalks, brown and dry) are still tall, but the new, vigorous, green segmented stems will be the tallest soon. The vegetation of the fen a thicket that stands as tall as me, a jumble of reeds, nettles, and the others. Further off, the osiers and willows on the river-bank, grey-green leaves. I sit under the same lone willow as a week or so ago, looking over the expanse of the fen. Its texture excites me – a layer of lines like needles or pins, the straight-up lines of the stems of the reeds, the perpendicular lines of their leaves, a greyish-green colour consonant with the willow's. A harmonic scene. Through this static but wafting texture floats a coot, and above it fly butterflies: peacocks, speckled woods, red admirals, weaving their way through the outstanding tallest stems that poke out through the top surface of the texture.

And the smell excites me as well. The scent of the meadowsweet sings like a state of mind. Their inflorescences hover above the rest of the vegetation, white galleon-clouds floating over the ocean. They generate a scent. Volatile molecules built from a repeated backbone of five carbon atoms (monoterpenes). Secreted by specialised cells in the flowers, they evaporate into the air. The scent attracts pollinating insects, and transforms my perception of the scene.

Just now I found some water lilies. In a sun-dappled pool beneath overhanging trees. White floating flowers. As if made of paper. Like origami. A sense of antiquity about them. Recent constructions of the ancestral tree of flowering plant species have identified the water lilies as occupying a branch that spilt away from the others long before most of the species and genera we recognise today even existed. Of course this doesn't mean that the water lilies I see in front of me are ancient – they are modern organisms that have been evolving for the same length of time as all the other creatures that are alive in the world today. But they represent an idea, a model of what the most primitive of flowering plants looked like.

I picnic on the seat that overlooks Wheatfen broad. Looking over common reeds, an expanse of water with ripple lines in it, the alders and willows at the further edge, and, above them, a grey-blue cloud hangs like a crag. There is distant thunder as if a stone rolls slowly down a scree-side in the sky. How do I observe? I describe what I see, the taste of the scene. I am sensitive. I receive what my senses sense. There's no impediment there. My mind moulds it in the reporting, the describing to itself of the observation. It's okay. Everything we see is refracted through the prism of mind. We bend what we see, because the bending allows us to see more. We test our next observations against the prediction of the bend. Just now I looked at water lilies, thought of the tree of life, felt that the flowers were in some way sacred, the thinking gave colour to the observation.

In St Mary's churchyard, on the grave, little has changed. A bit frustrating actually. I want to continue the story of the thale-cress. But the slowed growth means that there is little story to tell. The plant is wizened, drying. The skeleton-thin stem, with its four seed pods, is all that remains. The greenness of stem and leaf fading. Seed pods purple and buff. Three now shattered with seeds still clinging to the inner septum of the pod (seeds tiny, dark brown, ovoid), or seeds gone and septum retained as a thin finger. The plant is dying. Is there a moment, a threshold to cross, beyond which it is completely dead? I think it is more of a continuum than a line. But there is life in the seeds.

Tuesday 27th July

ON THE UBIQUITIN-PROTEASOME SYSTEM

To a string-quartet concert in the Mill at Hoxne, Suffolk. A barn in a wet field. The wind stirring the surrounding trees and moaning through the timbers of the barn, so that the four strings eased their way through that shifting band of sound like parallel wires through cheese. Then the weather calmed and a pair of bats flickered around the ceiling beams.

As the music played, my mind wandered. What is it that the seeds of the thale-cress plant in St Mary's face? Individually, an uncertain future. As a group, they have a chance. Life will continue. Recently I wrote about the tree of life. It's an appealing image. Perhaps it was like this: a drop of sea water and protein enclosed in a membrane. Progeny from that beginning then fanning out over billions of years so that each can be traced back to the first. There is universality in the idea. Shared heritage.

A commonality of working is predicted. Conservation of function. An example of this: the ubiquitin-proteasome system. Something that works in the cells of yeast, humans, and thale-cress. A mechanism for the selective destruction of proteins.

Proteins regulate life's processes. As enzymes that catalyse metabolic reactions or as transcription factors that drive the expression of particular genes during development. But these regulators are themselves regulated at many different levels. Regulated via control of the transcription (into mRNA) of the genes that encode them. Via control of the rate of translation of that mRNA into protein. Via control of the stability of the mRNA. Via control of the degradation of the protein itself. The ubiquitin-proteasome system is the main pathway for the selective degradation of proteins in multicellular organisms. It has two functions: a selecting/tagging function and a destruction one.

First, the selecting/tagging function. All multicellular organisms contain a small protein made of seventy-six amino acids and known as ubiquitin. This ubiquitin is a reusable tag. It becomes attached to specific target proteins by means of a group of enzymes that perform the attachment reaction in response to an activating signal. Once tagged, the target protein is recognised by the destruction function: a multi-subunit protein complex known as the proteasome. The tagged protein enters the proteasome and is then degraded by protein-digesting enzymes that are present in the proteasome interior. The ubiquitin protein is released undamaged and returns to find another protein to be tagged.

The ubiquitin-proteasome system is crucial to the regulation of life. It provides a way of controlling the levels of growth-regulatory proteins in cells. Whilst I was listening to Mozart, DELLAs were regulating the growth of the plants that surrounded me, the wet grass in the field, the trees moaning in the wind.

Previously, I described how gibberellin causes modification of DELLAs. This modification is an activating signal, a signal that causes a DELLA to be ubiquitinated, tagged, and thence destroyed in the proteasome. This is how DELLAs disappear. They disappear because they're selectively destroyed. Their destruction releases the growth of plants from the restraint they impose.

Wednesday 28th July

Today it's hot. The sun beating on my back as I cycled to work, the air like warm water.

Later, to Wheatfen. Just wonderful today. A thick-pile carpet of grey-green reeds, flecks of purple (the flowers of the willowherb and the purple loosestrife) that are in harmony with the green. These colours have a sweet and pleasing effect on the mind, I think via the evocation of memory: heather, honey, autumn rambles. The scent is sweet too – has a penetrating quality. The fen hums with life. Here is bindweed twirling around a reed. I see the genes and proteins that I study as part of the life of this fen. Further on I find the rosette of a dandelion of such exquisite radial symmetry growing out of the disintegrating wood of a rustic bridge of branches. And at the moment of finding and looking down I also see a horsefly on my calf and slap at it. The fly crumples and falls to the ground leaving a red smear of blood on my leg. That dandelion rosette hasn't flowered. Will it wait now for next summer?

Attracted by another tiny dandelion that is growing out of the trunk of a fallen tree. It has the tenderest of green leaves emerging out of spiky grass in a long trough of material that is the beginnings of earth: rotting fragments of bark with dust that has blown into the crevice. The trough like a stratum-layer of the log. A fresh dandelion. And there are others like it, a little younger, tiny and clinging to a small cleft that is higher up and parallel to the first. The plant that first caught my attention has a large chunk eaten out of one of its leaves. But I prefer the new ones. So tiny, such perfect miniatures. I try to move the leaves a little to see them more clearly. But they're very fragile, and I fear disturbing them. I can see two cotyledons already yellowing and two first true leaves. Exquisite, tender frailty. My story could have been about these. The building of these from cells, genes, and proteins, it's much the same as it is for the thale-cress plants. Differs in the details that make the dandelion a different thing, but otherwise much the same.

Thursday 29th July

THE SEEDS LIE DORMANT

It's now the height of summer. Hot, sky blue with a layer of haze. Lovely to cycle in a short-sleeved shirt.

We've completed the gravitropism experiment. As with the salt experiment, we got a positive result. The roots of normal thale-cress seedlings turn in rapid reorientation of their direction of growth after being moved. All quickly return to growing downwards. But DELLA-deficient roots are much less sure where to go. They do turn but much more slowly than the normal roots, often setting off first in the wrong direction. Eventually, they also grow downwards, but it takes them much longer to get there.

I think I can safely say that I am rekindled. My thought fecund again. In the space of less than a month I've had two separate new ideas, each of which will be the germ of a new research project. Working on/thinking of/writing about these new projects will be the major focus for the rest of this year. Along with whatever grows in St Mary's.

Lovely also to think about the thale-cress seeds lying on the soil in St Mary's. Perhaps in tiny depressions or crevices walled by particles of grit or sand. Lying dormant, progress arrested.

Waiting for the right conditions before they germinate. An extraordinary state: although superficially inert, metabolism continues within that dull brown coat.

Alice has been doing 'bug-hunting and mini-beasts' at school. She is fascinated by some galls growing on the leaves of the lime trees in the garden. These galls are projections that stand proud and perpendicular to the plane of the leaves like prehistoric menhirs standing in a field. They are maroon, distinct from the dark green of the leaves, created by the larvae of some gall-making wasp or other insect. What's remarkable about these galls is that, although made *by* the insect, they are made *of* the cells of the lime tree. Somehow the insect larva produces molecular signals that hijack the previously expected developmental trajectory of the leaf cells and forces their genes to send them in a different direction – to grow and divide out of rather than within plane, to make pigment, to build a structure they would otherwise not build. Astonishing plasticity.

Saturday 31st July

Seen from my study window, the beech in the garden next door is already hinting at the turn. Already, and it's still only July. There's a slight dinge, a tinged yellow quality to it. As if the colour were hovering, an aura surrounding leaves and branches. When does autumn begin and summer end? I'm a little unsettled by the sight of it. Of the tree preparing for winter. The summer secure no more.

AUGUST

Sunday 1st August

IT IS still, and high cloud covers the sky. Looking up into the oak from the study window, the branches seem silhouettes of green, not three-dimensional, immovably pasted on a flat grey layer of sky.

It is some time since it last rained. The garden feels dry: the euphorbias browning towers, growing from dust and dry-crumbed soil. How quickly the soft is replaced by the harsh. The seeds from the thale-cress plant in St Mary's won't germinate in this weather. I do hope that they *will* germinate eventually. It's possible that the mother plant was so weak that it gave inadequate support to their maturation. I do hope I'm wrong about this. That I *will*, sooner or later, see seedlings springing up. See thale-cress restored to the grave.

Tuesday 3rd August

It's getting to be hot. The sky cream-blue with haze. White clouds of slow-moving fluff. The sycamore has brown keys hanging on a background of dark green leaves (the August-green is darker, bluer-green than the yellow May-green). As I write I have the feeling that these notes should emphasise change. Progression and the process of it. Not be a series of static pictures.

To Wheatfen and St Mary's. There was a cooling breeze from the sea. On the way the sight of a patch of yellowing in a green tree tightened my chest like a shadow of anxiety. A presage of autumn/winter.

At Wheatfen the layer of green that is the reed-bed has a new thin crust on its surface: the brownish-purple of emerging flowers. Since I was last here, the flowering stems have pushed their way out of the tunnel of interwrapped leaf sheaths. They are grassy grasses, remind one of the maize field I cycled past on my way here.

The insects are wonderful. A humming buzz of dragonflies, weevils, butterflies, and beetles. Already a dragonfly with patches of shining aquamarine on its thorax and abdomen has settled on my notebook, and some kind of weevil has walked across the page.

Reeds, willow-herb, nettles in profusion. Luxuriant, rapid growth. Driven by release of the restraint of the DELLAs. But all doing it differently to generate their different characteristics and shapes. Butterflies today: lots of small tortoiseshells, peacocks, painted ladies, fewer red admirals than before. Some brimstones, speckled woods. Fewer meadowsweet flowers now. Guelder-rose berries turning from yellow to tinged orange.

Then to St Mary's churchyard. The two sides of its square walled by tall, dark chestnuts that cast full shade. Looking up into them: large, round, spiky fruit-cases hang with their weight. At the grave the skeleton of the thale-cress plant is dessicated, wiry, brown, and thin, almost gone. And no sign of new seedlings. Although I looked hard. Crawled around the grave, eyes close to the soil. But no surprise here. There really hasn't been enough rain.

One of those cycling thoughts on the way back. That we're obsessed with our own importance. We constantly present ourselves with an anthropocentric view of the world. Our literature focuses on human relationships in an urban setting. We depict for ourselves only a tiny fragment of the entire picture. Can we change this? If not, our centric view might destroy the whole. Yet we are what we are. We *will* see it all with ourselves at the centre. It's in our nature. So what are we to do?

Sunday 8th August

EVOLUTIONARY CHANGE AT THE LEVEL OF THE GENE

En route to Ireland. The last few days taken up with preparations for the journey. It is late at night, on the ferry to Cork. The boat heaving in the dying hours of a storm that crossed the Atlantic from west to east these last few days. I lie on my bunk, writing in my head, keeping my mind channelled in the hope of staving off seasickness.

I wonder what those seeds in St Mary's will do during the next three weeks or so whilst I'm away? And then there are the reed flowers in Wheatfen. Now so distant – on the east of England and me beyond the west of Wales. They are exquisite feathery things. The branch stems are green, but the tiny glumes (the flimsy membraneous flower-case covers) are wine-dark purple. The different shadings of green and purple give the whole flower head a lovely mottled appearance.

The purple is the colour of the pigment anthocyanin, the green areas regions that lack it. The pattern with which the anthocyanin is distributed is controlled in a way that is best understood in maize, a grass quite closely related to the reed. Maize controls its pigmentation pattern via a transcription factor that regulates the expression of pigment genes. This transcription factor is called BOOSTER (B), because it boosts the intensity of pigmentation in cells that contain it. Thus, the gene encoding B (*B*) is activated in those cells that are destined to become purple. B then activates genes encoding enzymes that perform a chain of reactions that results in the production of anthocyanin. Different varieties of maize have different patterns of pigmentation. For example, some have purple stems, some have purple leaves, whilst others have neither. Others display subtle differences in pigmentation of the glumes. This variation is due to variation in the sequence of the promoter region of *B*, and, perhaps surprisingly, not due to differences in the protein that *B* encodes.

These findings have important implications. Evolution works by change. The changes in the *B* gene could have been changes that alter the way in which B works (changes to the workings of the B protein itself), or changes to the time or place that the B protein is made. It seems that most changes to *B* involve changes to the promoter, the thing that controls expression of the gene, rather than changes to the protein that the gene encodes. Furthermore, if the difference between different strains of maize is the result of changes in the pattern of gene expression rather than of the protein that the gene encodes, then what about the differences between plant species, between maize and reed for instance? Perhaps the proteins made in each are largely the same, with the differences between the two being due to subtle differences of timing and cellular specificity of expression of those proteins. Is evolution less driven by change in protein sequence than by change in gene regulatory regions?

And another thing occurs to me as the boat churns and lurches. B is a transcription factor. A protein that controls the expression of genes by interacting with other promoters. *B*, the gene that encodes B, has its own promoter, and variation in the structure of that promoter alters *B*'s expression. Structural variation in the *B* promoter alters expression because it changes the way in which that promoter itself interacts with yet other transcription factors. So *B* is not a discreet entity. It is part of a complex network of genes that regulate other genes via the transcription factors that they encode. There is an image that popularly prevails of the gene being a simple unitary thing, a thing that does, on its own, a single, clearly defined task. This picture is too thin. Lacks the richness of the reality.

Monday 9th August – Ahakista

So here we are again, back once more in lovely Ireland. As always there's the feeling of being immersed in the climate. From the west, from across the bay, come shower and sun-patch, clouds of all shapes and densities, a momentary rainbow with ends stabbing the ground and describing an arc that spans the bay: all with such changes and volatility.

And of course it's green. The clichés have their validity. Green in a huge gradation of colour, as if there was a whole additional spectrum within the green division itself: emerald, chartreuse, jade; sea-green and sage; verdant from cedar-dark to the lemon-green of fertile pasture.

First, I'll have a few days rest. But then I must get back to thinking. About our two new projects: salt-growth response, gravitropism. I'm looking forward to it. I know that the thoughts will flow well here.

Wednesday 11th August

Today we walked out in heavy showers. Sometimes the best ideas – the newest ways of seeing the world – are the most fragile. Against this, the scientific view can sometimes seem so confident, so robust. But it's a fractured view. I'm struggling to see the world as a whole. So that it all makes sense. Is that possible?

Thursday 12th August

Last night I saw the universe in a new light. A concert in Bantry House. In the library, its huge windows giving on to the garden. The fading light replaced at the interval with a bright ring-chandelier of lit candles. The rain in the garden steady and vertical. Dancing and singing through that gentle, roaring rain was the exquisite fiddle of Martin Hayes. Accompanied on the guitar by Dennis Cahill. It was passion and restraint in simultaneity. They thumped the floor with their heels. And around the sparse predictability of rhythm wove yearnings: stretchings and squashings of time and pitch. The whole of life was in it.

Music works on many levels. There are the fundamental things: melody, cadence, harmony and so on. The ornamental and shadings. We perceive the layers simultaneously, make a totality from them. But the processes of life? So difficult to appreciate when the fundamentals are things that we can perceive only at a distance, through a microscope, via the logic of genetics, in terms of the interactions between molecules that we cannot see.

Saturday 14th August

A bright day. Mackerel sky. Our house a white-faced stone cube with slate roof, windows edged with blue. Nestling within a grove of pines that murmur in the breeze, a moan laced with liquid birdsong.

The house sits on the flank of a mountain ridge, the finger bone of the Sheep's Head peninsula. Above us, and behind, are gorse, rough olive-green and harsh brown grasses, patches of purple bell-heather: it's blanket bog and moorland, a thin layer topping stone. Below is Dunmanus Bay: today halfway between mat and gloss, unruffled grey-blue. Between us and the sea: gentle pastures, cultivated fields. In total it is a huge amphitheatre of landscape and weather.

Last night more music. Songs of love and loss; of life, death, and transience; of flowers and ground; rain and wind. Music, landscape, life, all connected.

Sunday 15th August

Walked down the road, then across mountain land to the old stone circle near the sea. Said to be around 3,000 years old. Eleven stones. Most of them fallen, but two still upright, standing on edge, one of them close to a wizened holly tree, branches wrapped around it like fingers grasping.

I thought of DELLAs, of how their structures are more ancient than this pattern of stones. Yet how a DELLA is individually transient. Each one destroyed after it is marked in response to gibberellin. That much (I think) is known. That's how plants grow.

There's a mechanism the plant has for detecting when the DELLA is marked, and for targeting it to the proteasome for destruction. This mechanism takes the form of a thing called the SCF complex. Three different proteins, symbolised by S, C, and

F, work in association. Most relevant here is the F protein. It has a special function. It recognises and binds to the DELLA, and is especially attracted to DELLAs that are marked. Once the DELLA is caught by the SCF complex it's tagged by a chain of ubiquitin markers – the ticket for entry into the proteasome. On arrival inside, the DELLA is destroyed, its component amino acids released (and perhaps then used in the construction of another protein). All these proteins – DELLA itself, all the SCF components – are transitory in their individual existences. Yet the things they do, their functions, are absolutely ancient.

Monday 16th August

Cool winds from the north bring clouds of varying density of grey. It rains in bursts: sometimes heavy globes of water cannon into the ground, sometimes pearly mists weave with less certain direction. Because of the clouds' volatility, the vector of the light is in constant flux. In seconds the looks of things change – blades of grass, flowers, berries, and leaves of hawthorn – texture becomes more or less pronounced, undersides more or less apparent. As the origin of the light shifts to a different place, so things are seen in new ways, and each new way is different again from the one before. All this was observed during a wet walk in the lane close to the house.

The lane is an avenue with hedges on either side. In the hedges: the fuschia that grows so robustly in the mild moistness of West Cork; the red flowers harmonise with the developing berries of the hawthorn; purple blackberries, orange Crocosmias, all tangled with hart's-tongue ferns. Further into the roadside the grassy verges are full of meadowsweet, their scent strong in the humid air between showers.

Tuesday 17th August

A lift east, then spent the day walking back west along the ridged backbone of the peninsula. Bantry Bay to the right, Dunmanus Bay to the left. Glorious sunshine, occasional clouds, one heavy shower sliding over the Beara peninsula but coming nowhere near to us. A sudden shade from a tiny cloud covers the sun and brings apprehension. Of what? Rain? Death? Perhaps both. And somehow the contact with the landscape momentarily balances simultaneous anxiety and comfort in the thought of death.

We picnicked at the peak of the ridge. The children happy in the sun. After eating, we read to one another. The ridge of the mountain is all rocky outcrops, heathers, and rough brown grasses. Smelling wonderful: of peat and heather flowers (both bell and ordinary heather). A vegetation all woven like a carpet that covers the rock and makes the peaty soil. Walking back along the ridge-top with the two headlands of the Mizzen and the Beara ahead of me on either side, I returned to thinking the thoughts I must soon get back to in earnest. About tropisms: how plants bend roots and stems to get the best from the world; a root round a stone then back into earth or towards water or nutrients in the soil; a stem to gain better access to the light; and how the bending and twisting of all these cylinders around one another makes the fabric of vegetation along which we so happily tramp westwards into the evening sun.

Wednesday 18th August

The holiday continues in this wonderful place. Our lovely white house. Dunmanus Bay spread out before us like a rumpled grey sheet. The moss-green part-wooded lump of the Mizzen peninsula beyond.

Later to Dereen gardens, on the north side of the Beara peninsula. Just into County Kerry. The gardens present a scene with classic resonance. A vista from the house across lawns, through trees to the sea below. A valley at the foot of an Olympian peak. Exotic, romantic plants growing in shady groves: eucalyptus and tree-ferns. Aromatic scents and oils in the moist air. Banks of rhododendrons. Such fertility amongst ooze. I had a strong sense of awe: of how the architectural majesty of the tree-ferns, their ancient structural pattern, is a consequence of the structure, the separation and replication of the two strands of the DNA molecule.

And then the gods on Olympus scraped a chair across the floor: thunder groaned and clapped, rain fell as if poured from a bucket. We ran across the lawn, nakedly exposed to this extremity of water, totally drenched in an instant. Minutes later, the sun came out again.

Afterwards we played on the beach, amongst floating yellow seaweed, clothes drying on our bodies in the sun. A lovely time looking into rock pools – the contained vibrancy of them: darting shrimps and feeling anemones. Tiny hermit crabs housed in winkle shells. I looked up from a pool, saw a flying heron, saw it momentarily as an image of the kind a painter might make – a few lines for beak and flapping wings and a cylindrical blotch that in some way represented, or was an abstraction of, the idea of 'heron, flying'. And would have represented the same to you had you seen it in your mind, I think. Does this way we have of seeing and then redrawing the world in our minds make us feel part of it? What would happen if we could do this with our scientific images: make a commonly understood sketch (or abstraction) that represents a DELLA, say? Would there then be a familiar, at-home sort of feeling about these things, which are as much part of our landscape as flying herons?

Thursday 19th August

It's strangely cold for August. Beams slanting down on to the Mizzen peninsula from a hole in the clouds – seen across Dunmanus Bay. As the hole moves, so the shape of the Mizzen is picked out differently. The same thing is transformed – folds in the landscape appear and then disappear. I find myself reflecting on the landscape of proteins. There's similarity here. I remember thinking something similar whilst in Wharfedale at Easter.

Friday 20th August

Still cold, and this morning a wind from the North brings an autumnal feel to the weather: there's a thrill in the tension between the chill of the breeze and the heat of the sun.

To Dooneen strand. Low tide – rock pools full of striped mollusc shells; red, green, and purple anemones, green and yellow seaweeds. A seal's glossy grey head bobs and watches us from a short distance out to sea. We hear him snorting and huffing. Lovely views back down Dunmanus Bay.

Saturday 21st August

The air remains cool. But today it's utterly still. The bay a plate-glass reflector of rocks, trees, and houses on the opposite shore.

I'm on holiday, and it's a time for reflection. How have things changed, what progress has there been since I began making these notes and sketches in January? Then, I was at a low ebb, having difficulty in seeing the connection between things. But now I see more clearly. I see that our science is relevant.

In particular, I am today in awe of the intricate complexity of the world. It's a commonplace to say that DNA and its replication have made the living world that we know. Although I don't think that this truth is really absorbed into the way that we see our lives. Even for me. Watching the milking this morning at the farm next door, I had to remind myself that black-and-white cows, cow dung, straw, the smell of milk and feed-nuts and shit, the flies, even the structure of the milking barn itself, are all the consequence of the replication of DNA.

And DNA is not the only thing that is essential to the scene. Take DELLAs for instance. Without DELLAs there would be no straw, no cows or cow-dung, no milking parlour, no us. Just one of the countless different molecules and structures, the necessary and interdependent parts of what we call life. Yet we cannot see these things. It's as if we were listening to a symphony and could only hear the sounds of highest pitch, not the bass note on which the harmony is built.

Sunday 22nd August

A heavy storm in the night. Awoke to the groan and rush of wind in the trees, the creaking of rafters, the banging of the sash-windows. The rain hurled like grit against the slates of the roof. And us snug in bed, the wood stove keeping us warm despite the weather. I lay there loving the noise of it all and the sense of security that the heightened contrast of outside and inside brought. I love the phenomenal, the sense of being in contact with the world that always seems so much stronger here than ever it does in Norwich. One is constantly aware of the slightest change in the weather, the gradation of wind speed, light or shade, rain. It makes time seem more real – marks it out somehow – so that each moment is felt, savoured, then gone. Unlike the more monotonous, less featured time in Norwich. The reality of life, death, transience is closer here.

This morning it's calm. No wind, sun between white clouds. The light strikes the mountain in yet another new way. Different features are brought into relief or enhanced by shadow.

I've begun to sketch out a possible shape for a paper on the gravitropism story. Although there's much still to do. I'm imagining several sections. The first shows that DELLA-deficient thale-cress mutants have an impaired gravitropic response. Which means a description of what we found towards the end of July. When normal seedlings are grown on agar in a Petri dish (the surface vertical, parallel with the force of gravity), the roots grow downwards, along the surface of the agar. When the plate is turned by ninety degrees, the roots quickly do a right-angled turn and then grow in the direction of the (new) down. But when the DELLA-deficient mutant is subjected to this treatment, its roots turn much more slowly.

The subsequent sections of the paper aren't yet clear. But they concern the biology of another plant hormone, a hormone called auxin, which I mentioned before in passing. It is auxin that causes the root to curve. When a growing root is moved by the turning of the plate, it's no longer in vertical orientation. Auxin accumulates in the cells of the (new) underside of the root. This accumulation inhibits the growth of lowerside but not the growth of upperside cells. The result is that the root grows in a curve. Our new observations suggest that DELLA-deficiency affects the auxin biology of roots. Either DELLA-deficiency prevents the occurrence of auxin accumulation, or it prevents the lowerside cells from responding to it. Our experiments will, in the next few months, enable us to distinguish between these possibilities. Whatever we find, it will be new.

Monday 23rd August

The qualities of the vegetation here influence the mind. There are rushes growing in thick clumps in a damp area of common land down the road from the house. They're green and brown, the flowers and some stems brown and the other stems green; and the sight of them, whenever my gaze falls there, plucks at a string, strikes a chord of melancholy. Whether the sound is to pervade my whole mind or whether it is to fade as soon as it is struck or when I look away is a different matter. But the thing is that the colour and texture of the rushes have that power or potential to move the mind. Why? Is it due to an association with some forgotten memory?

And other plants can do the same. This morning, whilst fetching wood from the shed, I glanced up at the sycamore behind the barn. Suddenly, I saw on it the first hints of autumn colour, a slight orange tinge to the green of some of the leaves, the buff-brown of the seed keys. The sight jerked at me. Like a missed heartbeat. It's an early sign of autumn.

The stream of thought is constantly diverted by sudden apprehensions. All the time the mind is pricked by changing temperatures, sun in the eyes or its sudden fading, wafts of scent, distracting noises. Yet our scientific thought – the understanding of the meristem, for example – takes place in front of that background and despite it. At any moment its train might be broken by the intrusion of the rest of the world. Perhaps this is what makes science so hard? Science concentrates the mind relentlessly on the subject in its view, to the exclusion of the rest of life. Insensitive concentration, separating science from the world. I think these notes are an attempt to address this contradiction.

Wednesday 25th August

I'm writing this at the top of the ridge behind the house, looking over Bantry Bay: Hungry Hill faces me, Seefin to my left, the slate of the bay flecked with white, the wind blowing the perfume of heather and gorse into my nose. The sun is warm, but the light and heat of it are weakened by the cool north wind.

I walked up over the three rises that lead to the top, the sun and wet ground making sweaty humidity in the shelter of each rise. Finally, past the old fields and deserted dwellings. Those fields scraped into existence by the fearful effort of forgotten hands. Still they endure, marked by the broken-down stone walls that edge them. We walked west out to the end of the peninsula yesterday, and it's the same all the way along: the remains of ancient dwellings, outlines of abandoned fields. Despite being slowly reclaimed by the bog, their character as fields still visibly retained.

Yesterday as I walked I realised that I think more now about death than I used to. Is it the same for everyone? When I was twenty, I never considered it, but now that I'm nearly fifty I do. Just sometimes, when I see a particular thing in a new way, say when the light catches the waves breaking at the end of the peninsula in a pattern I haven't noticed before. And I realise that that particular moment's perception, no matter how hard I try to capture it in words or to retain it in my memory, that unique vision will die with me.

Thursday 26th August

Today a walk over Seefin, a picnic at the summit, then on to the road down into Kilcrohane. Lovely the feeling of being pushed out into the sea by the land, with Bantry Bay and Dunmanus Bay on either side. The landscape moving and solemn. Especially the magnificent Hungry Hill, which rises so steeply out of the sea. It's truly picturesque, solid yet shifting in the ever-changing light. Why does it move us in this way? It occurred to me that every cell has a landscape that is similarly beautiful and should command similar reverence. Yet it does not. But it's hard to continue this line of thought since I don't entirely understand why the landscape affects us as it does. A cultural thing that started with the Romantic poets, perhaps? I'm not sure. The effect of a moving landscape is so visceral, seems innate.

I've been sketching what might go into the other paper, and it too is to do with the landscape. About how the world outside a plant shapes its growth and development. I think this paper will be an important one. For one thing, it explains why DELLAs exist at all. Previously this was a real puzzle: If DELLAs are so important to the life of plants, then how is it that plants lacking them are so similar to plants containing them? (There's little difference when the plants are grown in optimal conditions.) The new idea is that DELLAs help plants to respond to adverse conditions, real 'survival of the fittest' stuff.

Adverse conditions (too hot/cold, soil too salty, drought, etc.) are stressful, and it might be in the plant's interests to slow growth down and await better times. We showed that salt in the soil inhibits the growth of roots but is less inhibitory to the growth of roots lacking DELLAs. The next thing we need to look at is survival. To test if the slowed growth conferred by DELLAs in adverse conditions enables plants to survive those conditions. My guess is that it does. That toxic salt levels stabilise DELLAs, that the resultant slowed growth enables plants to 'sit out' the stressful conditions. So that when conditions improve, growth can be resumed. This is something we need to test.

Friday 27th August

Today we went in a boat to see the seals on some rocky islands in the bay. In the shallows there was a shoal of sprats spangling in the sun. Sudden unanimous shifts in the direction of a thousand or more tiny fish seen as merged flashes. Made me think about DELLAs, how what we describe in the 'relief of restraint' model talks about one protein but what actually happens is mass action of many molecules, a shoal of proteins.

Monday 30th August – Oxford

The last few days travelling home – a calm crossing on the ferry, staying a couple of days in Oxford *en route*. This afternoon walked to Binsey Church. And to a shock. Last time we were here there was a graceful avenue of high, luxuriant horse chestnuts which made a tunnel and funnelled the walk into the churchyard. But now it's gone. Felled, all felled. Why? My first sight of the stumps a sudden blow like a fist in the eye. The scene is diminished, bereft, lacking. And why here, of all places, where it might be thought that 'Binsey Poplars' would hold some currency? Not one remains, just two parallel lines of severed stumps.

Why? Fear of litigation apparently. The risk that a branch might fall and injure someone. So are we now to fell every tree in the land? Do we carry on trying to bend the world, to mend it, render it safer until we have no world at all?

SEPTEMBER

Thursday 2nd September

BACK in Norwich. Back home. An early autumn picture in the study window. And the light has autumnal quality. Each day the sun is a little lower in the sky. The light slants.

Today the air is still, with tiny movement only. The leaves of the lime are static. Each stays in one space with no wind to move it to another. And the colours are changing. The beech leaves are the colour of diluted orange juice, whilst on the lime a small bunch of leaves is seen from the window as a splash, a fleck of custard-yellow on the background of sombre green.

In me there is a tingling excitement. It's a special feeling that always comes with autumn. What is it? Why? Is it from hearing the roar of life more clearly now we're reaching the edge of winter?

Yesterday I went to St Mary's. Expecting by now to see some thale-cress seedlings. But still there is nothing. What's happening amongst the grains of soil on the surface of the grave? When will those seedlings appear? I saw the tiny brown seeds clinging to the septum of the burst pods. Then, a few days later, they were gone. This was about six weeks ago. What are they doing now? They must be somewhere. The soil is damp, from dew and from recent rain. Surely they'll germinate soon? I want to see that next generation. The next phase of the cycle's turn. In my excited state their absence is a needling frustration.

Friday 3rd September

An utterly still morning. In early autumn sunlight that's on the edge of gold. It elicits feelings I find hard to express: a kind of reverence/joy/wonder mix. Today's light, the whole scene before my study window, stimulates. Once, the prevailing view was that the world was made by God. A God unseen but manifest in Nature. Now, as a result of scientific advance we tend to see Nature differently. But it's often said that there's no mystery to the current view, or that the mystery is somehow tarnished. As if our deeper understanding somehow diminishes awe. But I don't accept this. The world is glorious, as glorious now as it ever was.

And then there's the question of purpose. Looking at the natural world, at the structures of plants and animals, it's natural to think that there is a design or pattern underlying the mechanics. Indeed, science itself often uses machine imagery in descriptions of organisms, representations of signalling pathways, and so on. We frequently compare organisms to the structures we ourselves have designed and built to fulfil a purpose. These ideas of purpose have their origins in the view that God designed the world.

Science has revealed a different narrative. That there is no grand design, no power checking that all is going according to plan. We change by random mutation. If that change results accidentally in an improvement in function, it persists. This new narrative is perhaps less comforting, less attractive than the old. But there is still space for appreciation. I think we should assert the continuing appropriateness of awe. Find a way of leaving behind the idea of divine purpose without simultaneously abandoning a sense of wonder.

Tuesday 7th September

Surlingham glorious in warm golden sunshine.

The chestnuts are quite magnificent now. Leaves still green hands but aged (brown and yellow patches are clear) by the progress of the year. Their time is nearly up. And hanging, pendulous, the spiky globes of the nut-cases. Green sputniks in clusters, waiting for the drop to earth soon to come.

But best of all, I think, *I think*, that I've had my first sight of the new generation of thale-cress seedlings. Not easy to see though. Only when I went down on my hands and knees, scanned the corner of the grave with my magnifying glass. They are totally delightful, exquisite jewels shining through the lens. Green cotyledons. Vivid, with a slight tinge of mustard. Bright against the black wet soil. Surely these are the progeny of the wispy skeleton they surround. There's connection here. The next generation. The cycle turning.

I did a quick sketch of some of them. There are fourteen seedlings growing within a rough four-inch-diameter circular area of soil that surrounds the shrivelled stem. The seedlings aren't scattered

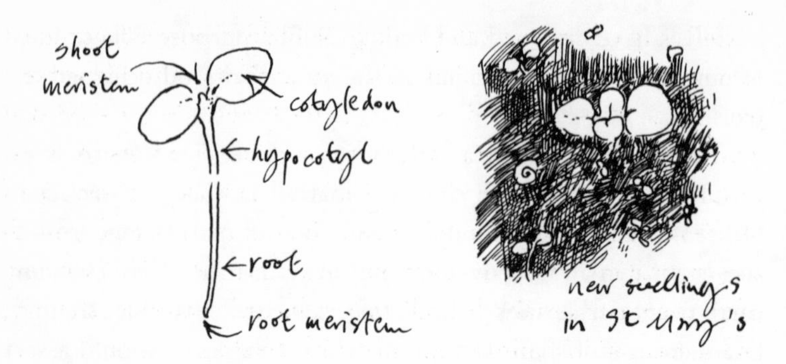

Thale-cress seedlings. *Left*, a diagram of a single seedling. *Right*, some of the new seedlings.

uniformly but lie concentrated in one area as it were in a cloud, like a rain-shadow thrown by the now-dead mother.

A more precise description of an individual seedling: it grows by an unfolding and expansion of the embryo that was in the seed. What a remarkable transformation. From tiny, near-invisible, dry, inert dot to fleshy seedling. All in a couple of days. Each seedling a pair of cotyledons in opposing orientation (the shoot meristem lying between them), a hypocotyl, and a root that's already penetrating the soil (although this latter is of course invisible to me). The structure is as it was in the embryo, but magnified. The hypocotyl grows by the expansion of its cells. A huge increase in size, each cell expanding to a hundred or more times its original volume, each a cylinder blown long by the pressure of turgor and the relaxation of the restraining walls. DELLAs control this, the growth of the hypocotyl cells being due to the regulated release of DELLA-mediated growth restraint. And whilst the hypocotyl grows, so do the cotyledons and root. Here both cell expansion and cell division play a part.

The result of all this: the seedlings. Tiny crucifixes planted in the grave.

DELLAs co-ordinate the timing of thale-cress-seed germination with climatic variation. This co-ordination is important, because an ungerminated seed is relatively safe. But a seed that germinates is chancing its hand. Playing its cards against feckless fortune. Within the seed is a survival package. A ration of nutrients stored in the cotyledons. These nutrients fuel growth as germination proceeds. But the ration is finite. The seedling must eventually establish itself as a viable self-dependent entity, and it has ration-limited time in which to do so.

Finding these seedlings is a reassurance. There's a sense of old certainties being restored. The comforting swing of the pendulum. Seed to seed to seed, the ticking clock of the years.

We found a year or so ago that one of the five thale-cress DELLAs plays its own specific part in triggering the germination of thale-cress seeds in response to water. And now I'm reminded of this I can't imagine why it took me so long to think of our recent new research direction. It was staring us in the face. We already had an instance of a DELLA being an interface between the inside and outside world of the plant: a DELLA that regulates seed germination in response to the environment. Why didn't we move more easily from this particular to the more general view?

Wednesday 8th September

ON RGL2 AND GERMINATION

Yet another bright beautiful morning. As if one could see the concentric isobars described around the cyclone that floats in the atmosphere above our heads. Running in the still air I startled two wood pigeons in my path. I hadn't seen them until the very moment they began to move. It made me think of how the universe is composed of things we can see and things we can't see, but that sometimes a change in the quality of invisible things can make them visible.

To return to the germination of seeds. To the discovery I mentioned yesterday. Gibberellin-deficient thale-cress seeds don't germinate unless they're supplied with gibberellin. Previously I described how the dwarfed stem growth of gibberellin-deficient mutants can be restored either by gibberellin application or by removing the DELLAs GAI and RGA from the cells of the plant. But removal of GAI and RGA doesn't make a gibberellin-deficient plant completely normal. Seed germination still requires the application of gibberellin. This suggested that, whilst GAI and RGA control stem growth, one or more of the other DELLAs (RGL1, RGL2, or RGL3) regulate seed germination.

We tested this prediction quite recently. And what we found has the beauty of order.

In the years following the cloning of *GAI* there were many great advances in understanding of the biology of thale-cress. One of the most important, as I've mentioned before, was the sequencing of the thale-cress genome. At around the time that the genome was being sequenced, several labs established collections of thale-cress lines carrying transposon insertions (many thousands of independent ones) for which the particular gene sequence interrupted by the transposon was known. It was possible that there might be mutant lines carrying transposon insertions in *RGL1, RGL2*, or *RGL3* within one of these collections. To our great excitement, we found some, lines carrying insertions in *RGL1* and *RGL2*.

These insertions disrupted the structure of the genes that contained them, preventing those genes from working. So we looked to see if we could detect any differences between plants carrying these new mutations and normal plants. At first the plants lacking RGL1 or RGL2 looked just like normal ones. The seeds germinated, the seedlings grew, the adult plants flowered with flowers whose structures were just like those of the non-mutant plants growing next to them. But plants lacking GAI and RGA also look relatively normal, and it's only in plants with reduced levels of gibberellin that the effects of a lack of GAI and RGA become clear. So we crossed the mutants lacking RGL1 or RGL2 with the *ga1-3* mutant (the mutant that is gibberellin-deficient) and obtained plants lacking one or other of these DELLAs and lacking gibberellin as well.

Now came our unexpected discovery. We planted some seeds that lacked RGL2 and also lacked gibberellin. Just in case there would be something to see. We didn't really think there would be. The seeds were gibberellin-deficient, after all, and probably wouldn't germinate. A few days later, to our astonishment, they grew: we could see seedlings: squat hypocotyls holding green cotyledons flat to the sky and soil. Germination *had* occurred, despite the lack of gibberellin. And as we watched these plants over the following weeks, it became clear that their subsequent growth was exactly like that of gibberellin-deficient plants having normal levels of RGL2. They were dark green and dwarfed, bore flowers with short petals and stamens.

So RGL2 is especially involved in controlling the germination of thale-cress seeds. RGL2 does something that the other DELLAs do not. And the implication of this observation is that it was RGL2 itself that was blocking the germination of gibberellin-deficient seeds. That the gibberellin in normal seeds overcomes the effects of RGL2, thus permitting germination.

We found that RGL2 regulates germination in a very special way. The seeds that are shed from the mother plant are desiccated, containing only a little water. They need water to germinate. Furthermore, water triggers germination. When a dry seed is exposed to water, it soaks it up. As this happens, germination begins. In addition, there's a rapid rise in the levels of *RGL2* messenger RNA (the RNA transcript that encodes RGL2) in the cells of the embryo. This rise causes an increase in the level of RGL2, and RGL2 blocks the germination of seeds. Thus water sets the process of germination in motion, and at the same time puts a block to that process in place.

Now thale-cress seeds don't merely require water to germinate. They also need light. Thale-cress seeds don't germinate in the dark. When the seeds are exposed to light, that light promotes the production of gibberellin. Gibberellin then causes a reduction in the level of *RGL2* mRNA and promotes the destruction of the RGL2 protein. This removes the block to germination imposed by RGL2, and permits germination to proceed to the next stage.

So this is what happened to one of the seeds on the surface of the grave in St Mary's churchyard. There's something exquisite about the sensitivity of it. The rain came. It drenched the soil. The seed soaked up the water, swelled like a sponge. *RGL2* mRNA levels rose in response to the water. RGL2 was made, and blocked further germination. But this was a temporary block. A checkpoint. The seed was on the surface of the soil. Sunlight penetrated the seed-coat and stimulated the cells of the embryo to make gibberellin. This gibberellin reduced the level of *RGL2* mRNA, caused the destruction of RGL2. Thus the block was removed, and the seed progressed to the making of a seedling.

It's a series of checks and balances. And there are others. They ensure progression through to full germination, but only if the right conditions prevail. Both water and light are needed. There is harmony in this.

Saturday 11th September

The forecast is for more rain. The weather maps show the rings and bullseye of a deep depression crossing the Atlantic and aimed at southern Scotland. At its edge is rain – and that rain is due here tomorrow.

Sunday 12th September

ON THE GROWTH OF THE HYPOCOTYL

Overnight there was a storm of strong wind and pelting rain. The trees boiling with it, the noise a constant roar. Rain in thick drops, of a volume that's rarely seen. It was like this for hours, and the ground is drenched, sodden. This is really good for the seedlings, will enable establishment. And this morning the Indian summer returned.

Morning sun then mist. Cycling down to the river on the way to work, I saw the sun slide from rich yolk to pale cherry-pink tomato as I plunged into a wall of mist.

Germination mostly involves the expansion of cells. As water is absorbed, the embryo begins to expand until it bursts out of the seed-coat. Once out, growth continues. The growth of the hypocotyl is particularly fascinating. Its orientation and extent are a function of the genes within the cells of the hypocotyl, the rest of the world, and the sun beyond.

The sun is a globe in the sky that makes light that reaches earth. That's how we usually see it. But where is the edge of that globe? Think of the sun as a nucleus of a larger globe defined by the spread of its light. Then we are all part of the sun and the germination of seeds is a property of the sun itself.

Monday 13th September

A glorious, golden late summer morning.

Cycled down Bracondale, sun full in my face, real vision often only at the rim of sight and that misty. The sudden knowledge, first from feeling, not from conscious hearing, that the distant bell of City Hall had begun striking 8.00. Only with the second stroke could I be sure. Yet I'd known it – known that the first stroke had been sounded even though not aware of hearing it. It was felt. The second stroke travelling through the milky, still air.

And then on to Surlingham. Shocked by the sudden whine and Doppler shift of a motorcycle. Its rider clad in mat black leather. Weaving between me and a car coming from the opposite direction, banking steeply. I shivered. I once saw a rider like this lying bent in the grass on the side of the road. A robot-jerk quality to movements directed unconsciously or from the edge of consciousness. His hand at a sinister angle to his arm, wrist broken at the very least. I saw him as jelly held together by the leather that encased him. I don't know if he lived or died.

Then to Wheatfen. I sit under the willow and look out over the fen. There is a sense of wonder in the wetness. The fen more fabric than before. The reeds still stand up, but between their parallel lines there's now more filling – more tiller stems of reeds I think. But also, woven amongst them, weft to the warp, the stems of bindweed, nettle, willow-herb, purple loosestrife – all winding and binding into the cloth of it.

Still some butterflies – red admirals, small tortoiseshells, but definitely fewer than before. Flying over the purple surface of the fen are paired and coupling dragonflies.

The flowers of the reeds are more opened now, less compact than when last seen, each purple floret unfurled. Whilst in the more sheltered woody part of the fen, I tapped one and saw, for a brief instant before its dissolution in the wind, a shimmering cloud of pollen. Peering closely at the florets I could see the tiny pale-yellow anthers from which that pollen comes. In the sun-filled, open fen the seed flowers are further on. More feathery, more dissected, brown rather than purple, no longer shedding pollen.

The light coming more from behind than above allows a clear vision of the colours of the fen. Although the green of the reeds still predominates, I've the feeling that it's less uniform than before. Some leaves are beginning to turn brown, tapering to brown tips, some are shredded by the wind with buff edges to the shreds, others have traces of yellow or purple.

There's the buzz of insects, the chirruping of crickets. What is it about this time that's so enthralling? The light, yes, the high sky of duck-egg blue, the layering of the distant clouds, these too. But perhaps also the knowledge that it's fading as I watch.

Further round the fen – the meadowsweet heads are now thick and furry clumps of seeds, each one a tiny black or brown dart with hairy feathers attached. Thousands of them on each head. And these heads emit no fragrance now. They are dry and dying. But what potency there is within the seeds that they carry!

On to St Mary's. Horse chestnuts in a majestic ring, orange leaves against the blue sky. To the grave. Seedlings bright in the sun. But one of them, I now see, is in the shade of a thin-fleshed new dandelion leaf. This shaded seedling has a long hypocotyl. Is taller, thinner than those in full sunlight. Close by, a

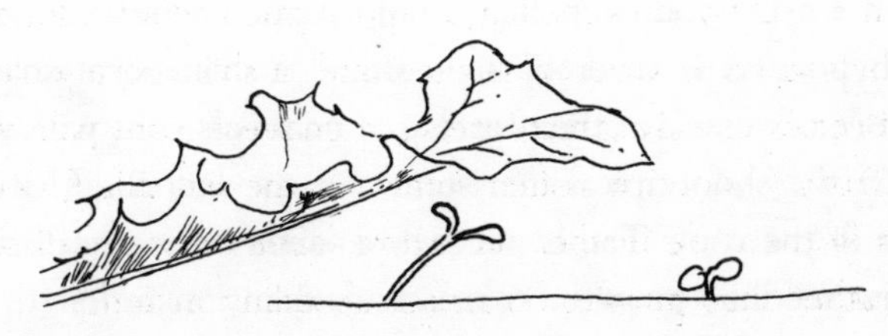

The thale-cress seedling that is shaded by the dandelion leaf is taller than a neighbouring unshaded seedling.

horse chestnut case thumps the ground. And that thump makes me feel small with respect to the grandeur of the scene. I'm aware of ascents and descents. The fall of that chestnut case. The rise of the cotyledons at the top of those expanding hypocotyls. The split-on-impact that reveals the oiled brown conker swathed in white, fleshy membranes.

The cell walls of the hypocotyl relaxing and reshaping in response to the pressure within. This moment will endure in my memory.

Humility isn't part of the lexicon of modern science. Nor is wonder. We don't write about these feelings in our papers. We are in a double-bind. To admit of wonder is to become personally involved. Personal involvement, feeling, clouds objectivity. Objective is what we're supposed to be. Whilst wonder is what really drives us, and wonder is what we feel, we cannot admit of it. Little surprise, then, that non-scientists often do not understand us. That the picture we scientists present can sometimes seem evasive, to lack real feeling, to be without pith. It's because a core foundation of the picture remains obscured, actively repressed.

Part of my wonder at this present scene is in the shaded hypocotyl being longer than those in the open. This difference is the product of an interaction between the internal workings of the cells of the hypocotyl and the quality of the light within which it grows. It's a phenomenon that's only partially understood.

The hypocotyl is an embryonic stem. A cylindrical structure with a core of vessels at the centre that connects root with shoot. At its top the shoot-apical meristem and cotyledons. At its base the shaft of the root. The hypocotyl expands during germination and initial seedling growth. A massive expansion in length from seed embryo to final seedling hypocotyl. All achieved by the longitudinal expansion of cells.

The extent and orientation of hypocotyl expansion is determined by the qualities of light. By the intensity of the light, the direction from which it comes, its spectral properties (the relative amounts of the different wavelengths of which it is composed). Thus the sun, its centre so many million miles away, is controlling the growth of these particular hypocotyls. They can see the light. Their cells contain proteins which act as light receptors. These receptors are remarkable molecules. They detect the presence of light, then communicate that information to the genes in the nuclei of cells, altering gene activity in a way that alters the growth of the hypocotyl. Like me, these seedlings see and respond. As for me, the chain of events that leads from vision to response begins with molecules which absorb light.

The light detected by the photoreceptors inhibits hypocotyl growth. This is part of a process that's familiar to us all. Plants grown in the dark are pale, tall, and spindly. They grow with exaggerated expansion. It's an adaptation that promotes survival – the plant needs light to survive, and it will only find light if it grows itself quickly out of dark places. So a plant in the dark arrests the expansion of leaves (useless since there's no light to collect) and promotes the expansion of stems (increasing the chance of pushing itself out of the dark). In the light, the growth of stems and hypocotyls is inhibited, the expansion of leaves promoted.

The geneticist's approach to understanding light-dependent hypocotyl growth inhibition is to look for mutants in which the inhibition by light is reduced. To search for seedlings which have tall (rather than short) hypocotyls in the light. It's easy. And fun. Spotting that one amongst thousands, its cotyledons raised high above the canopy of the rest, a tall tree amongst a forest of short ones, is very exciting.

Finding a long-hypocotyl mutant amongst shorter normal seedlings.

A mutant hypocotyl that grows tall in the light (long-hypocotyl) is impaired in its ability to detect or respond to that light. The thale-cress genome contains a handful of genes which, when inactivated by mutation, result in light-insensitive expansion of the hypocotyl. Amongst these are genes that encode a family of light-receptor proteins, proteins known as phytochromes.

Phytochromes have astonishing properties. They are amazingly subtle in the way they work. So sensitively discriminative of the quality of light. For they don't just respond to light, they respond to particular colours, particular wavelengths of the light spectrum. Evolution has driven them in the direction of this exquisite colour-sensitivity.

Phytochromes are proteins of labile shape. But I haven't got time for this. Now I have to organise myself. Tomorrow I fly to New Zealand, the beginnings of a two-and-a-half-week trip to New Zealand, Australia, and Singapore. Although I'm excited to be going, I also resent being separated from things here.

Tuesday 14th September – Singapore

An hour in the airport, in transit between London and Sydney, a relief to be out of the aeroplane. Behind me, incongruously within the marble-faced concrete, is a little shingled and glass-walled triangular oasis, bursting with luxuriant tree-ferns and the ground-hugging rosettes of other ferns that look like our own heart's-tongues. The smell is sweet, so calming after thirteen hours of aeroplane air.

I'm tired but feeling pretty good. Rather enjoying the excitement of it. I'm about to fly over the equator (for the first time). And it's wonderful to have the opportunity to read, to read without interruption. I'm rereading William Golding's *Sea Trilogy*, his fictional record of an early nineteenth-century voyage from England to the Antipodes. My appreciation heightened by the huge contrasts between the journey it describes and mine. There is so much life in there. And the ferns have restored me.

The purpose of my journey is twofold. First, to meet with collaborators in New Zealand. Second, to give a 'plenary' talk at a large international conference in Australia.

Wednesday 15th September

ON PHYTOCHROMES

I slept on the plane to New Zealand. And whilst I slept, I dreamed. Of another still day in Norfolk. That I was cycling to work in weak sunshine, sky feeling so high, the only cloud a flat layer of haze in the uppermost levels. In the valley bottom there was a layer of mist like milk in a bowl, a surface flatness to it into which I plunged on my bike. A sudden chill on bare arms (shirt-sleeves rolled back). Looking up – the sky still blue, but with added greyness and grain-like smoke.

When I awoke I returned to thinking about the grave in St

Mary's, the thale-cress seedlings on its surface. To the phytochromes that connect those seedlings to the sun. And the idea that the seedlings are themselves a layer of the sun. The thinking somehow appropriate to my situation: life sustained within a bubble of steel thousands of feet above the earth's surface.

Phytochromes are proteins to which a small molecule known as the chromophore is attached. The linkage of protein to chromophore generates a structure having a fantastic property. This structure can absorb photons – units of light energy. It is this property, this ability to capture the energy of sunlight, that makes the phytochrome a light receptor. The structure detects light and then signals this to the rest of the cell. In the elongating hypocotyl, detection of light by the phytochrome results in the inhibition of cellular expansion. That is why mutants lacking phytochromes have a long hypocotyl when grown in the light.

Phytochromes can exist in two different states, an 'active' state (known as P_{FR}) and an 'inactive' state (known as P_R). These different states have different structures, and thus different identities. Change from one state to the other is triggered by the absorption of a photon of light. But it is richer than this. The phytochrome distinguishes between the different wavelengths of the light spectrum, and has different affinities for those wavelengths depending on whether it is in the active or inactive state.

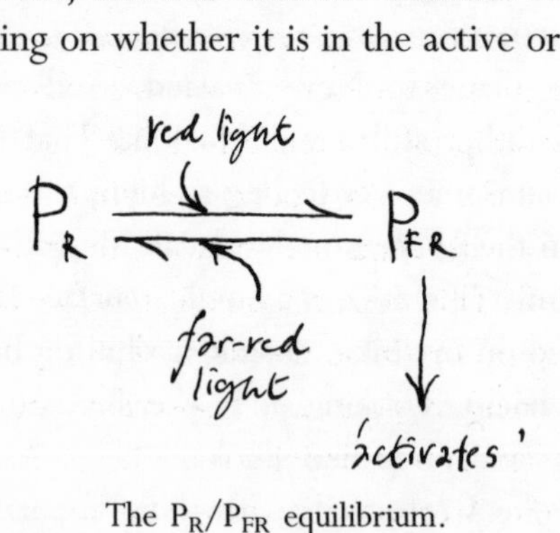

The P_R/P_{FR} equilibrium.

P_R selectively absorbs red light, which converts it to the active P_{FR} state. Conversely, P_{FR} preferentially absorbs far-red light, causing it to revert to the P_R state.

When exposed to sunlight, the phytochrome molecules in the cell are constantly flipping backwards and forwards from one state to the other. The relative proportions of active to inactive state reflects the relative proportions of red to far-red light in the sunlight spectrum.

So what is the point of all this? How does it help the plant to control its growth? Only very recently have some of the answers to these questions been found. Once activated, the phytochrome moves into the nucleus of cells. This was shown by studying phytochrome-GFP fusions (in which the phytochrome is fused with the green fluorescent protein that I described previously).

In plant cells grown in the dark the phytochrome is in the inactive state and fluorescence (due to GFP) is uniformly distributed throughout the cytoplasm of the cell. But when these cells are exposed to a pulse of red light, the now activated phytochrome (seen as fluorescence) rapidly concentrates in the nucleus. A stimulating finding. The genes are in the nucleus of course, and now the activated phytochrome is in the nucleus as well, making the interaction of genes and phytochrome possible.

Further experiments revealed what happens when the activated phytochrome enters the nucleus. It forms a complex with another protein, a transcription factor called PIF3 (for phytochrome-interacting factor 3). The phytochrome–PIF3 complex then binds to DNA, to a specific sequence found in the promoters of light-regulated genes. This binding modulates the activity of these genes, speeding up or (in some cases) slowing down the rate at which their coding regions are transcribed into mRNA, thus regulating the levels and activities of the proteins those genes encode. It is these proteins that change the growth of the plant in

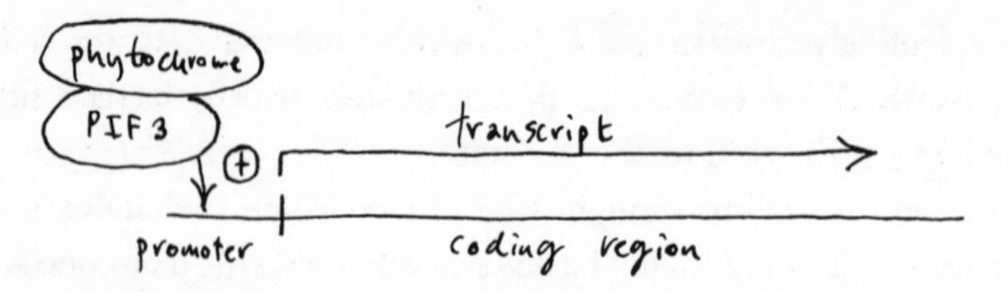

Gene activation by phytochrome–PIF3. The complex of activated phytochrome and PIF3 protein binds to the promoters of genes, where it activates transcription (the making of the mRNA copy) of the gene coding region.

response to light, and cause the light-induced inhibition of hypocotyl growth. No doubt DELLAs are involved in this process, although the details are not yet known.

The connectivity of all this is thrilling. There's a chain that connects the sun to the growth of those seedlings on the grave in St Mary's churchyard in faraway Norfolk: the sun emits light, the light falls on the seedlings, the phytochrome is activated, moves from cytoplasm to nucleus, complexes with PIF3, the phytochrome–PIF3 complex binds to selected gene promoters, gene expression is changed, plant growth is altered, seedling hypocotyl growth is inhibited.

The fact that the new seedlings in St Mary's have hypocotyls of different lengths, that the one growing beneath the dandelion leaf is taller than the others, can be explained by the shading of the leaf: less light, less activation of phytochrome, less inhibition of growth. But there's more to it than that. When the sunlight hits the dandelion leaf, some of that light is reflected, some absorbed; the remainder is transmitted through the leaf and so reaches the seedling below. Of course this shaded seedling is seeing less light and growing taller as a result. But the light it is seeing has been altered both in intensity and in composition by its passage through the dandelion leaf. That leaf is a filter. Its cells absorb some wavelengths more than they absorb others. In particular, red light is absorbed more than far-red light. So the light that filters through the dandelion leaf and reaches the seedling beneath it is enriched in far-red light. That far-red light inactivates the activated phytochrome. Less activated phytochrome means reduced inhibition of hypocotyl growth means taller hypocotyls.

It's a breathtaking example of the power of selective force. The phytochrome system a mechanism that promotes elongation of plant parts that are shaded by other plants. Enabling growth from shade into sunlight. Improving access to the light on which continuation of the life-cycle depends. It's a system that is exquisitely adapted to what's needed – a specific ability to sense those changes in light quality that are precisely characteristic of the shade cast by other plants.

Friday 17th September – Palmerston North, New Zealand
A striking landscape. But I'm tired, and sense of distance prevents that landscape from impinging at the moment. As I rested I found myself remembering a day about this time last year in Norfolk. With the children. A day of light and shade. Sunlight and showers. Small clouds of heavy rain, wind moving them on, stacking them; one by one they drenched the ground so that it steamed in the intervening hot sun.

We were blackberrying. Briars rough with berries. Some green, some red, some purple, some rich, fat, and shining black. Some already shrivelled and dried, bluebottles basking on them. Wasps. Purple stains on my fingers. And the memory of purple takes me further back. To heather, walking on the Yorkshire moors with my father. Years ago. A blustery afternoon; clouds low, grey, dour, imposing. I, a child of eight or nine. My father and I sheltered in a shooting butt. We had walked many miles across the bleak brown landscape. Along tracks of millstone grit, peat, and sand. Now and again a startled grouse rose with raucous cries.

At the start of the walk I'd been excited. To go alone with my father on such an adventure. Had been receptive to the scene, the moors, the weather, the colours and smells of the heather and dying bracken. Had revelled in the challenge of walking against the wind. But we'd walked a long time and I'd felt my mind detaching from the outside world. I became introspective, noticed that my feet were sore. Anxiety was mounting: was the walk much further? Was I strong enough? A general irritability. Then, as the path crossed a line of shooting butts, my father said we should shelter in one for a while and eat our lunch. He wrapped me in a raincoat against the wind and passed me a sandwich. I bit into it. Felt the sharp contraction of muscles around my salivary glands. The crumbs of bread tasted sweet.

For a while I was aware of little other than the sweetness in my mouth. But then, I began to feel a lightening in my mood. Opening up to our surroundings, I could see the landscape of the green valley laid out below us through holes in the wall of the shooting butt. The images became ever sharper. More defined. I was interested. Followed the lanes in the valley from house to house, made connections.

I looked at the fields below, marked out by dry-stone walls, and they became countries apart, each containing their own lives. I was reminded of the story in the Bible where Jesus views the kingdoms of the earth from a high mountain-top. The sweetness was now in my mind, and I was happy, taking pleasure in the buffeting of the wind, the austere grandeur of the moors, the green warmth of the valley below. From sun and earth and rain to wheat, from wheat to flour, flour to bread, starch to sugar to tongue, gut and brain, from tongue, gut, and brain to mind. Our daily bread.

Sunday 19th September – Palmerston North, New Zealand

I've been here for discussions with colleagues about ways to collaborate on DELLA projects.

And during the last few days my excitement about being here has been growing. That excitement enhanced by a sense of dislocation: that I've been catapalted from early autumn into spring. There are lambs everywhere, and daffodils. From my window the gaunt, bare branches of towering deciduous trees. The sparseness of everything a particular shock since a few days ago I was experiencing the luxuriance of late summer.

Yet the journey does enhance the idea of the world being a single thing. Last night I had the sudden thought that our most basic approaches to seeing the world, our mathematics for example, involve separating one part away from the rest. We count starting with the number 1, and that act immediately separates the thing counted from everything else. Surely all this splitting and fractionating makes us see the world in one way and not in another. The closer we observe – the more we focus on something in isolation – the less we see the world as a whole.

And of course I do wonder how the new thale-cress seedlings are progressing. Their tiny roots penetrating the soil, inching closer to me here in New Zealand.

Tuesday 21st September – Canberra

ON PHOTOTROPISM

Thale-cress seedlings detect light via several different types of photoreceptor. Besides phytochromes, another class of photoreceptor is the phototropins. Phototropins enable plants to turn towards the light.

Phototropins sit within the external membranes of cells. Like phytochromes, they are light-absorbing molecules. Phototropin consists of two separate regions: a light-absorbing one and a signalling one. The light-absorbing region has a special affinity for blue light, just as a phytochrome has a special affinity for red light. Blue light activates the phototropin molecule.

The signalling region of phototropin has the properties of an enzyme, an enzyme that phosphorylates. This means that it has the ability to add a phosphate group (a molecule consisting of a phosphorus atom and several oxygen atoms) to the amino-acid components of proteins, in particular to serine or threonine. When blue light is absorbed by the light-sensing region of a phototropin, it changes shape. This change activates the signalling region, causing phosphorylation of serine or threonine residues, both within the phototropin itself (autophosphorylation) and probably within other (unknown) proteins as well. The subsequent steps in this chain of events are unknown. But what is known is that the sensing of blue light by a phototropin is responsible for the phenomenon of phototropism: the bending of plant organs towards light. We know this from studies of thalecress mutants that lack phototropins. When normal seedlings are placed in a dark box with a pin-prick in the wall, their hypocotyls bend towards the light coming from the hole. It's an adaptive

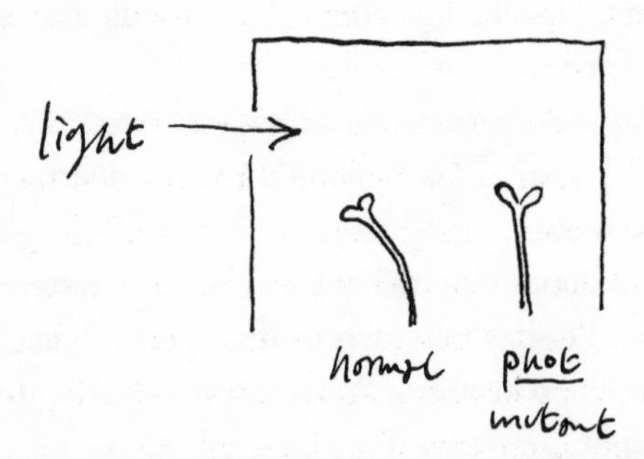

The normal seedling bends towards the light. The *phot* mutant fails to bend towards the light because it lacks the phototropin photoreceptor.

response. Selection has favoured seedlings that exhibit this response because it enables them to find the light they need for growth. Mutants that lack phototropins don't do it. Their hypocotyls continue to grow straight, they cannot see the light they seek.

Sunday 26th September – Singapore

Singapore has a mercurial climate. Hot. Sunshine, then abrupt wild rainstorms. Heavy globes of fast-descending water, jagged lines of lightning. The weather reflects my mind: volatile with circadian disruption. There is a parallel with the *Sea Trilogy* – the bizarre, disorientated state of the people on that stinking, creaking ship, the wild, drugged party on the world's spinning edge.

Yesterday to the rainforest. Steaming heat. Huge green spreading leaves. Shade. Such energy – sounded in whistling. In a screeching pitched at different levels. Above these sounds, woven amongst them, the bobbling calls of birds mostly not seen but hidden in the canopy. Huge insects – ants, bees the size of a baby's fist. Butterflies – orange and blue. Dragonflies with coppery delta-shaped wings. The feeling that it's all one thing, a light- and heat-speeded, water-driven torrent of life. A living steam-engine of flesh.

The interdependence of living things is so vividly apparent in the rainforest. The jumble of plants thrusting and tangling around one another; climbers and creepers, penetrating, grasping, embracing. A reflection of the connectivities within the plants themselves. Connectivities within the cells. That's what the science of our group has been about lately. Our big breakthrough in 2003 was the discovery of an internal connection. Something we'd suspected must be there although we didn't know what it was.

We'd previously reached an explanation for the growth of

plants that involved DELLAs. But the explanation was only partial. We knew that there are other things in plants that affect the rate at which they grow. Internal things, hormones other than gibberellin for instance. We didn't know if or how these other things related to DELLAs. In short, we didn't know if the DELLA system was truly fundamental to the control of growth.

We'd puzzled over this for a long time. The problem was a particularly difficult one. Its importance was clear, yet the way to solve it was not. We could see no way forward. No clear and unambiguous experimental test of the hypothesis that DELLAs are connected to anything else. Yet we had the feeling that they must be. That DELLAs couldn't have evolved unless they were connected to the rest of the biology of plants.

Then I had an idea. I'd been reading a book describing an experiment performed in the 1950s. An experiment involving gibberellin-deficient dwarf-mutant pea plants. Normally, these plants respond rapidly to gibberellin. If gibberellin is given to them, their stems rapidly begin to elongate. The region responsible for this elongation being the stem internodes that are a short distance below the shoot meristem. The experiment I read about showed something unexpected: that if the shoot meristem is removed from such plants, the gibberellin no longer stimulates the elongation of those stem internodes. Thus, gibberellin-regulated stem growth requires the presence of the shoot meristem, even though it is not the meristem itself that actually does the growing.

The hormone known as auxin flows down the stem of a plant from top to root-tip. Most of this auxin has its source in the meristem, and removing the meristem reduces the flow. Was it this reduced auxin flow that dampened the response of the stem internodes to gibberellin? Further experiments with

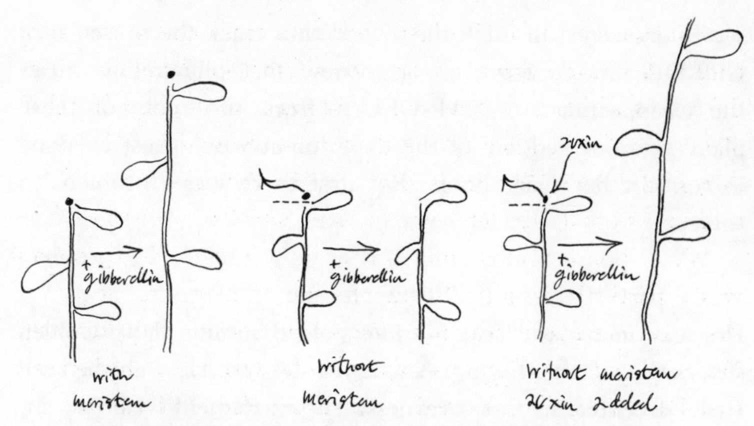

Gibberellin stimulates the growth of intact dwarf pea shoots (*left*), but not that of shoots lacking the shoot meristem (*middle*; meristem represented by black dot). Gibberellin response is restored if the shoot meristem is replaced with auxin (*right*).

dwarf pea plants were able to answer this question. When purified auxin was applied to the site from which the meristem had been removed, the stem-internode growth response to gibberellin was restored. Thus meristem-derived auxin permits gibberellin to stimulate growth.

Although I'd known of this experiment for many years, it was only whilst rereading the book describing it that its significance really struck me. The experiment had been done nearly half a century ago, at a time when nothing was known about DELLAs or about how gibberellin promotes the growth of plants by overcoming DELLA-restraint. I realised that we should redo this experiment. Redo it in a way that allowed us to look directly at the behaviour of DELLAs. We had the essential material with which to do this. The plants expressing GFP-DELLA. The roots of these plants would be a powerful experimental model because, being colourless, it is easy to see fluorescence through them, and thus track the presence of GFP-DELLA. And we already knew that gibberellin causes the disappearance of GFP-DELLA from the roots of these plants. So our redoing of the experiment would best be done in roots rather than shoots. But first there was a problem to solve.

Wednesday 29th September – Norfolk

Home again at last. To a glorious golden autumn light. Golden slivers, thin shafts. Fading. A thing to cling to. Dew on the reed leaves accentuating their greyness. The excitement is here again. Moments of sunbeam-sharp intensity.

In St Mary's churchyard there are conkers on the ground, some still part-covered by their smashed spiky cases. In the wild part behind the church there is a thicket. Tall stems of grass amongst the graves. Nettles. Cobwebs. Thick bramble vines twisted around ancient stone crucifixes. Plump black fruits hanging from them in trusses.

The thale-cress seedlings grew fast whilst I was away. And there are more of them, more seeds have germinated. Between the opposing cotyledons of the most advanced of the seedlings the first true leaves are clearly visible and expanding fast. The meristem is beginning to generate the leaf spiral that will build a rosette. Soon the beginnings of the next leaf will appear. The others are less advanced in their development, and some, those most recently germinated, are at the stage of expanding cotyledons only. Doubtless the roots are burrowing in, penetrating the soil with their tips.

The growth of roots. The problem we needed to solve. Although we knew that gibberellin causes the disappearance of GFP-DELLA from roots, we didn't know if it actually promotes their growth. We needed the answer to this question before we could use roots to determine the relationship between auxin and DELLAs in the regulation of root growth. So we did a simple experiment. We compared the seedling root of a gibberellin-deficient thale-cress mutant with that of a normal seedling.

We found that gibberellin-deficient seedling roots are shorter than normal seedling roots and that normal growth can be restored to mutant seedling roots if they're given gibberellin. So yes, gibberellin does control the growth of roots. Not surprising perhaps. But all that followed was dependent on this.

If gibberellin controls the growth of roots, does it do so by opposing the effects of the DELLAs GAI and RGA? Just like it does in shoots? To answer this question, we did more experiments. We found that the roots of gibberellin-deficient seedlings lacking GAI

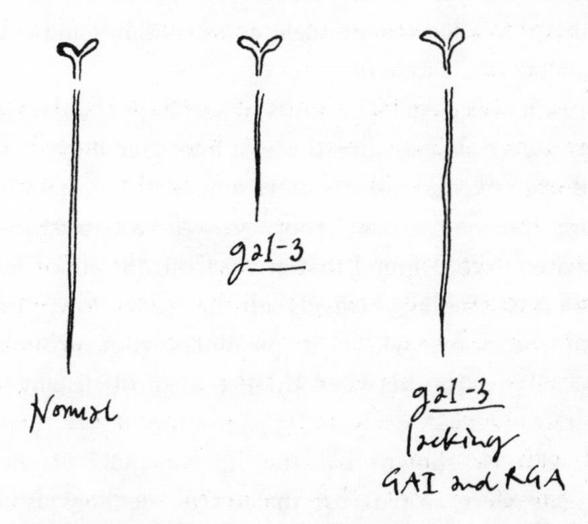

Gibberellin deficient *ga1-3* seedling roots are shorter than normal roots, whilst lack of GAI and RGA restores *ga1-3* roots to normal length.

and RGA are as long as those of normal plants. So gibberellin promotes the growth of roots, by overcoming the growth restraint due to the DELLAs GAI and RGA. Once again, this wasn't surprising. Stems grow from cells generated in the shoot meristem. But stem growth is actually driven primarily by the expansion and division of cells in stem internodes below that meristem. Similarly, whilst roots grow from cells generated in the root meristem, root growth is itself mostly the result of cell expansions and divisions in the elongation zone that lies just above the root meristem. Therefore, the fact that there is common regulation of these phenomena, that both are controlled by DELLAs, is not unexpected. But we needed to know.

Then what of the stream of auxin? Does this also control the growth of the root? The theory said that it should. But, so far as we were aware, nobody had actually tested it. So we did a very simple thing. We took some thale-cress seedlings and cut off the tops. Cut out the shoot meristem.

The result was clear. The roots of seedlings that lacked shoot meristems were shorter than those of intact seedlings. And if we replaced the excised shoot meristem with a drop of fluid containing purified auxin, root growth was restored. We were excited. We'd found that auxin from the shoot meristem of thale-cress seedlings travels all the way down the stem and into the root, where it promotes root growth. That there's a connection between these two most distant poles of the plant.

We had now shown that the fundamentals of the 1950s experiments were as true for thale-cress seedlings as they are for pea shoots. That both gibberellin and auxin promote root growth. So we could proceed with the replication of the original experiment, but this time using thale-cress roots.

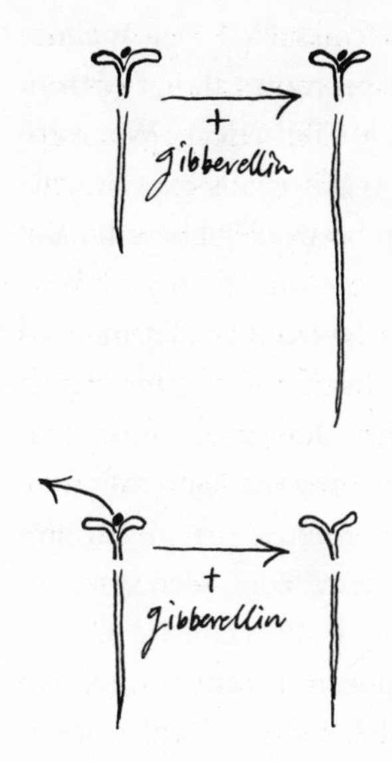

We began by growing some gibberellin-deficient mutant thale-cress seedlings. As before, the roots of these seedlings were dwarfed but grew longer when given gibberellin. Our experiment hinged on this root-growth response. What would happen to the growth of roots of gibberellin-deficient seedlings that lacked a shoot meristem? Would they still grow longer once gibberellin was given to them?

We removed the shoot meristem from some gibberellin-deficient seedlings, left them to grow for a few days, then gave them gibberellin. Over the next few days we kept returning to look at the growth of the roots. And gradually we started to see something: whilst gibberellin still promoted the growth of these roots, it did so much less than in the roots of intact gibberellin-deficient plants. And, if the excised shoot meristem was replaced with a drop of auxin solution, the growth of roots in response to gibberellin was restored.

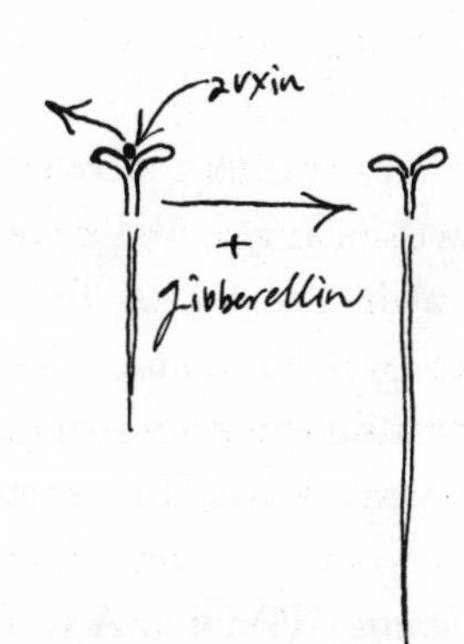

Gibberellin stimulates the growth of gibberellin-deficient thale-cress seedling roots (*top*), but not that of seedlings lacking the shoot meristem (*middle*). Gibberellin response is restored if the shoot meristem is replaced with auxin (*bottom*).

So the growth of thale-cress roots is controlled in a manner similar to that of pea stems. Gibberellin promotes the growth of both but requires the presence of auxin for full effect. We were excited to have been able to expand upon the 1950s experiment. To have shown that there is a relationship between gibberellin and auxin in the regulation of growth of more than one species of plant, that this relationship exists in roots as well as shoots. But most of all we were excited because we could now turn our attention to the role of the DELLAs in all this. Were they involved?

Our next experiment was a simple but revealing elaboration on the disappearance of GFP-DELLA in response to gibberellin treatment. To remove the shoot meristem from seedlings expressing GFP-DELLA and then determine if this removal altered the rate at which the GFP-DELLA disappeared from the cells of the root in response to gibberellin. Finally, to see if replacement of the meristem with auxin solution would restore normal behaviour to GFP-DELLA.

We thought it likely that we'd see an effect. It made sense that stopping the flow of auxin would stabilise the DELLAs, increase DELLA-restraint, and thus slow the growth of roots. We'd be able to see this in our experiment because the fluorescence generated by GFP-DELLA would no longer be dimmed in response to gibberellin. On the contrary, that fluorescence would persist in the root-cell nuclei of gibberellin-treated seedlings that lacked the shoot meristem.

The results of this experiment were exciting. GFP-DELLA was gone from the root-cell nuclei of intact seedlings within four hours of the start of the gibberellin treatment. But in seedlings that lacked a shoot meristem, those nuclei continued to glow. And in seedlings in which the shoot meristem had been replaced by a drop of auxin solution, the GFP fluorescence was gone.

These observations allowed us to add a new layer to our

understanding of plant growth. We had already found that growth is restrained by DELLAs. We had then added the finding that gibberellin promotes growth by stimulating destruction of DELLAs. We now knew that auxin potentiates gibberellin-stimulated DELLA-destruction.

We'd done it. Found a link that connected the gibberellin–DELLA growth-regulatory system to auxin, to something else we also knew to be crucial to the regulation of the growth of plants. It was a step towards the unification of understanding. And it's likely that this connection operates not only in roots but in shoots, leaves and flowers as well. And not just in the roots, shoots, leaves, and flowers of the thale-cress but in other plants too. The reeds in the fen. The chestnut trees. Plants everywhere.

This, then, was the discovery we published amidst much excitement in the spring of 2003. It was a new way of seeing growth, and I was awed by what we could now see, felt so lucky to have been a part of it. But the thing that came to frustrate me, in autumn/spring 2003/4, although I didn't fully realise it until a few months ago, was that whilst our discovery was revealing of the connections between the hidden, internal things that drive the growth of a plant, it told us nothing about how those things are themselves connected to the world outside that plant.

Now, however, we've moved on. The last few months of experiments have ended this frustration. And we have, in the last few weeks, made huge progress.

OCTOBER

Saturday 2nd October

Autumn advances. The air is cooler. The sun lower in the sky. The texture of things picked out in sharpened relief, shadows lengthened by the slope of the light.

Science is faddy, prone to fashion just like the rest of the world, flexes about looking for the next big thing. Important advances can lie hidden for years because other discoveries take the limelight. Until true value is recognised, via rediscovery, or duplication by scientists unaware of the original research. It can take years for true perspective to develop, before the major advances can really be appreciated, the ones that fundamentally change our way of seeing the world. And it is also inherent in the scientific process that things once established are often revised. Or swept aside by newer visions.

Sunday 3rd October

The orange spreads in expanding patches and lumps. On the oak there are clumps of yellowing, browning leaves: limp hands hang next to others that are still green. There are similar clumps on the lime; but these are darker brown, more desiccated and crumpled, look dead. And yesterday I saw a rowan that seemed to be one whole shape of orange-red. There's a sense of fast progression. That the patches of colour are coalescing at speed.

The wind stirs the trees, and the light moves with them. The sky pale blue, a solitary white cloud in a tower. There's a storm coming from the west, so the forecast says. There will be wind and heavy rain in the night. But for now the sunlight bounces from the white wall of our bedroom, through my eyes and into my mind.

Later, to St Mary's. The image of roots growing into soil is so exquisite. I remember writing about this in the spring. But, because of our recent gravitropism experiments, I know so much more about it now than I did then.

The rate and direction of root growth is affected by many things. Gravity plays a crucial role. Now, as a result of our most recent experiments, I can describe the mechanism of gravitropism in a more subtle, more profound way that involves the dense, starch-filled organelles known as amyloplasts and the way they shift when the position of a root is changed.

How does the shift in position of the amyloplasts cause a change in the direction of root growth? We already knew that auxin is involved. Auxin flows down through the central vasculature of the shoot and root to the root tip. Here, it fans back upon itself like a fountain, flowing back up the root through the outer layers of cortex cells and surrounding epidermis. In this way it reaches the elongation zone of the root, the region where the root grows by cell expansion (rather than division). And it is in this elongation zone that auxin controls the rate at which roots grow, by modulating the rate of cell expansion.

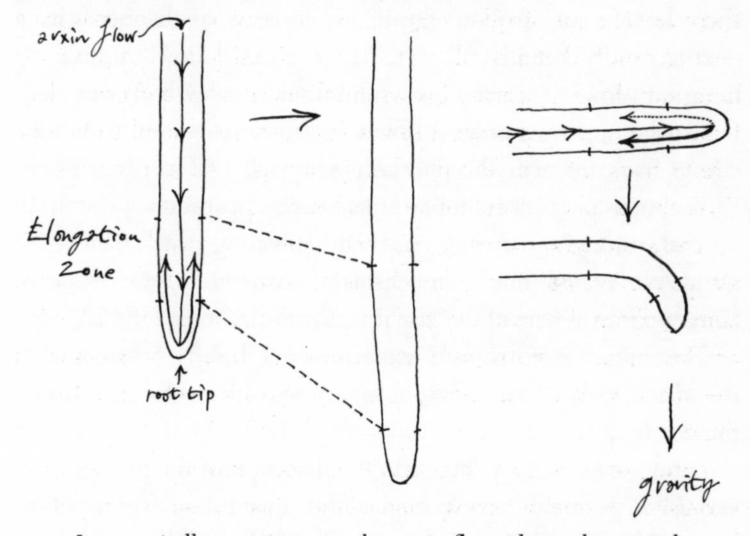

In a vertically growing root the auxin flows down the central vasculature of the root, then flows back in equal amounts in the peripheral root tissues on all sides, causing equal growth on all sides (*left*). When a root is moved into a horizontal position the reverse flow of auxin is diverted so that more auxin flows into the lower side versus the upper side of the root. The result is inhibition of growth of lower side relative to upper side, causing root curvature (*right*).

The upwards flow of auxin is uniform where a root is growing directly down towards the pull of gravity. However, when the root is shifted, the amyloplasts are displaced, and the stream of auxin is diverted. Exactly how amyloplast displacement causes this change remains unclear, but it is known that auxin flow to the (new) upper surface of the root is reduced, whilst that to the (new) lower surface is increased.

At normal concentration, auxin promotes the growth of roots. But at high concentration auxin actually has an inhibitory effect, and slows root growth down. This is what happens in a gravitationally stimulated root. The increased flow of auxin to the (new) lower surface causes inhibition of growth on that side, whilst the upper surface continues to grow. As a result, the root curves back towards the pull of gravity.

We had previously found that auxin promotes growth in normal, straight-growing roots by affecting the stability of DELLAs. Was a similar mechanism involved in gravitropism? Since auxin was part of the gravitropic motor, were DELLAs too?

The initial gravitropism experiment I briefly mentioned a month or so ago was a simple test of this idea. Our conclusion then was that thale-cress roots need DELLAs if they are to respond normally to a change in the direction of the gravitational vector. Presumably, we thought, the accumulation of auxin in lower-side elongation-zone cells of gravitationally stimulated normal roots causes DELLAs to inhibit the expansion of those cells. Whilst in a root lacking most of the DELLAs this cannot occur.

But then we realised that there were alternative explanations. After all, we had no evidence that the patterns of auxin flow are the same in normal and in DELLA-deficient plants. Perhaps DELLAs themselves affect the flow of auxin. We needed to test things further. And there was another puzzle. Previously, we'd shown that the basic, straight-line growth of a root is stimulated by auxin from the shoot meristem. But now, in the case of the curving root, we were proposing that the auxin that accumulates in the lower-side elongation-zone cells inhibits growth by inhibiting gibberellin-driven DELLA-destruction. How could this be? How could auxin both stimulate and inhibit the destruction of DELLAs? Right now, at the very moment of my writing this, we're running experiments that will throw some light on these questions.

This, then, is what the roots of those new thale-cress seedlings in St Mary's are doing. Growing down into the damp, cool earth. Slowly pushing on. Held to their path by the mechanisms I've just described. As if tied to a plumb-line by the sequential action of gravitational stimulus-sensing amyloplasts, an alteration in the reverse flow of auxin, a modulation of DELLA properties (perhaps . . . we'll know this soon) and resultant orientations of growth. Any deviations from the path restored to the true line in this way.

Monday 4th October

ON SPECIES

As predicted, there was heavy rain in the night. This morning the world has moved on a little. It's colder. A chill wind penetrates my jumper. The sky a cymbal-clash of lead and gold. Showers of brass rain.

Yesterday, we went to pick blackberries. Alice and Jack racing to fill their punnets. Blue-smudged faces. The scratchings of the thorns. The shining purple berries. The gathering in of the fruit, a harvest ritual.

Whilst picking I reflected on the way that much biological thought divides the living world into species. For instance, we group plants into sets. There is the set we call oaks, the set we call limes, the set we call daisies. We box them, focus on their differences, despite that fact that they have more in common than they do that divides them. Do these boxes hinder us from seeing life as one thing?

The lime in our garden lost many of its leaves in the night. There's a thin layer of them lying on the ground beneath the tree. The seeds remain hanging: dangling, pendulous buff spheres on stalks, shaking in the breeze.

Tuesday 5th October

Is there anything wrong with seeing the world as sacred? This morning I saw it this way. Without ambiguity. Felt the relief that comes from the acceptance of this view. And an associated warmth from being part of the world, not a cool, detached observer. But is this just a comfortable illusion? An easy answer? Now there's another voice in my mind pushing me in the opposite direction. Saying that distance is necessary for focus, for perspective. How to reconcile these opposites? And are they necessarily opposites?

Wednesday 6th October

The autumn advances. Darkness eats into the days. Okay, so we *are* approaching winter. But the world will continue to turn. I *am* approaching death. But life will go on. What, then, is there to fear? I'll attempt to maintain some serenity in the face of the coming winter. And that is easier to do within the context of the world as sacred.

Alice is part of my life going on. She has her own ideas about DELLAs. We sometimes talk about them whilst I walk her to school. She wants me to use them to slow the growth of a neighbouring sycamore that increasingly shades our garden!

Friday 8th October

ON DELLAS AND ADVERSITY

Yes, autumn is advancing. When I get up in the morning, the light is faint. Each day the evening dark comes earlier. Now, in mid-morning, there is yellow-orange sunlight of flattened incidence.

Stillness. Thin, flat clouds hovering in the sky. A nebulous crust of mist on the fen as I cycle to work.

Yesterday, we got some wonderful new results. Two different sets of observations at once! The latest developments in our salt story.

Our initial discovery predicted that the growth-restraining function of DELLAs somehow becomes more prominent in adversity. And the first of our two new experiments tested that prediction. Yesterday, we found that the fluorescence due to GFP-DELLA is much more intense in the roots of seedlings grown on a medium containing salt than it is in the roots of seedlings grown without it. So yes, the growth-restraining function of DELLAs *does* become more prominent in adversity. DELLAs accumulate in adverse conditions. The resultant increase in activity restrains growth.

But why should this be? Why should plants grown in adversity actively restrain their own growth? The second experiment addressed this question. And we've found that DELLA-deficient mutants don't just grow faster in high-salt conditions, but die sooner too. This observation brings insight. Now I begin to see why plants have DELLAs. Until now there was a question hovering somewhere not too distant in the back of my mind. What is the point of a mechanism to regulate growth if plants that lack it grow relatively normally? Now I think I have an answer to this question.

DELLAs aren't essential to growth itself but provide the plant with a way of tuning their growth to environmental conditions. This was not obvious to me before, because, in the lab, plants are usually grown in 'ideal' conditions. Getting out into the world has opened my eyes, helped me to think of an experiment that shows that DELLAs have 'adaptive significance'. That they enable plants to adapt their physiology to the vagaries of the environment in which they grow, and that plants having DELLAs are therefore more likely to survive such vagaries than are plants lacking them.

Now my mind is fermenting questions. Perhaps the most important is: how do the DELLAs know? What is it that makes them accumulate in response to adversity? This is the next thing we need to test.

The biology of adversity concerns those delicate thale-cress seedlings in St Mary's churchyard. The autumn approaches. It's getting colder. The rain periodically floods, then drought follows. Yet the DELLA-restraint system provides constant shelter. Shields those seedlings, protects them from oscillating climatic challenge. Speeds or slows growth as appropriate to the prevailing conditions.

Saturday 9th October

For a while today I was uneasy. With a passing fear that something might be wrong. Science attempts to represent reality. And here I am proposing that DELLAs are fundamental to the relationship between the growth of plants and the environment they grow in. That DELLAs promote survival in nature. A big claim. Could I be wrong? I've no reason to doubt, yet doubt creeps in. It's strange how one fluctuates.

How one can spring so rapidly from belief to doubt. Yesterday I was excited, positive, mind leaping with the implications of our new finding, ideas nascent. Today I'm uncertain, questioning. Both states of mind are I think a necessary part of being a scientist. Yet it can be a disconcertingly vertiginous existence.

Sunday 10th October

To Wheatfen. Low cloud, weak sun. Sky revealed in patches. An orange tinge pervades the vegetation. A tang to the taste, as it were. Hinted at everywhere, sometimes magnified in spreading foci.

There's a mild but persistent westerly wind. It hisses and roars in the canopy above the wood. Dislodging leaves which fall in sad descent. Looking out from the path to the fen, a guelder rose can be seen in a gap through trees, a brilliant red beacon of leaves and berries, an island of fire in the wet.

Sitting once more under the willow, it's clear that the water level is rising. The leaves of my notebook are ruffled by the wind, the butterflies are gone. The reeds are bent, perhaps by the storm of a few days ago. Their colours are changed again. Flowering heads buff-brown, no longer purple. And they've become feathery: what were so recently florets now contain seeds with long, silky hairs attached, fluffed up by the wind. The seeds themselves: tiny dark-brown points at the ends of the hairs. These changes are astonishing, have been so rapid. And the leaves have changed too: yellowing, browning, the green receding.

Whilst walking around the fen, the thrill of the wind reminds me of a concert we went to last autumn. Vivaldi in St Andrew's Hall in Norwich. Lovely string-playing against the banging and rolling of the wind in the rafters. The music expressive of the dramatic, of passionate flow. The art of the spectacular. Sudden changes in dynamic. The Picturesque. An evoked vision of a dusky Italianate landscape, crags and trees in a glowing Arcadia. Far-off clouds shining in the sun. A distant murmur of thunder adding transient threat to the pervading ease.

Somehow, although I cannot define it, there's an affinity between these conjured responses to a musical representation of the world and the things I'm trying to capture in these notes. A joy in phenomena? And there is more, because the music exists simultaneously on different levels, on different scales of focus, is like life itself in this sense.

Then to St Mary's. The wall of trees that surrounds the graveyard is taking on its autumn colours. Leaves edged with bronze. Branches hung with splitting chestnut cases, brown nuts peeping. The ground beneath studded with conkers. The thalecress seedlings are doing well – growing fast in the mild air. I'm conscious of an incongruity: of how the freshness of these seedlings contrasts with the fading of autumn, the seasonal disintegration.

Tuesday 12th October

Yesterday I took a day off and cycled to Reepham along the old railway track. Bright sunshine there and back. On the outskirts of Norwich, industrial developments are seen filtered through the trees that line the route. Dank canals, rank smells from a pharmaceuticals factory. Then on past Drayton and Thorpe Marriot into a more rural landscape. Woods separated by wide fields. Grey-brown stubble reflecting the heat of the sun into my face. Wetlands and reeds the mile or so before Reepham. Lunch, a pint, then home again.

All along the route the oranges and browns layering themselves, encroaching on the space that was green. Most of all the reds, the wild hips and haws. Berry-points of brilliance that coalesce into a single mass of colour, entire vermillion bushes, when seen from a distance. It is sublime. Exhilarating. Beneath all this an awareness of the coming darkness somehow accentuates the thrill.

I began to think about the way in which we see the location of the centre of the universe. Before Copernicus it was the earth. After Copernicus it was the sun. Now, who knows where it lies? But perhaps that centre is conceptually important. Perhaps it defines things that are central to the way we think. And I fleetingly wondered if we shouldn't revert to the older vision, return to seeing the earth as the centre. It is after all the centre of *our* universe. The stars are distant.

Monday 18th October

DELLAS REGULATE microRNAs

The colours are increasingly vivid. This morning, cycling beneath a sky of misty haze, I saw leaves that were pink, others cherry-red, some brown, some orange.

And there's something about autumn that makes it pungent with memory. Yesterday evening, in the gathering dark, I went for a run. There was a particular moment when I saw a distant figure and a street-light flickering between the shifting leaves of a tree and there was something about the totality of that instant, the things I saw combined with the smell of the decaying leaves, that jerked my memory into a picture of me as a child, holding my mother's hand one autumn evening, in short trousers, with scabs on my knees.

I've neglected these notes for the last few days. Quite simply too much to do. In my work diary for today: write grant proposal, review manuscript, write papers. Yes, I'm still thinking about those salt-growth and gravitropism papers. They're not coming easily.

But today I've been thinking about something else. Something we've been doing in parallel with the salt and gravitropism projects. We've made an exciting new discovery which relates DELLAs to a recent revolution in biological understanding.

In general, genes act by being transcribed into messenger RNA, by that mRNA being translated into protein, by that protein having a function. One way of controlling the activity of a gene is to regulate the amount of its respective mRNA in the cell. Low levels of mRNA result in low levels of protein, whilst higher levels of mRNA result in higher levels of protein.

The recent revolution is the discovery of a new class of RNAs, known as microRNAs, which control mRNA amounts. MicroRNAs are very short pieces of RNA, only twenty or twenty-one nucleotides in length, that are encoded by genes (genes that can actually be thought of as anti-genes) whose existence has only very recently been revealed. These microRNAs are complementary to a region of sequence from within the (much longer) mRNAs that encode specific proteins. Complementary means opposite, in the sense that the sequence of the microRNA can bind by base-pairing (in a manner analogous to the base-pairing that holds together the double strand of the DNA molecule) to the sequence of the mRNA. The result is a long single-stranded mRNA molecule with a short double-stranded segment (microRNA bound to mRNA) somewhere in the middle. Double-stranded RNA (unlike double-stranded DNA) is highly unstable, because the cells contain an enzyme complex that destroys it. This complex is thought to have evolved as a defence mechanism. It enables cells to destroy infecting viruses, many of which are in the form of double-stranded RNA. But the mechanism also provides for the regulation of genes. The evolution of microRNAs complementary to plant mRNAs enables destruction of those mRNAs, and hence the regulation of mRNA amounts.

In our case, we had recently read of the existence in thale-cress of a microRNA (known as miR159) that has near-perfect complementarity to the sequence of an mRNA known to encode a protein called GAMYB. GAMYB is a transcription factor: known to bind DNA at a particular sequence in the promoter of gibberellin-regulated genes, known to potentiate the expression of genes (the transcription of mRNA) as a result of that binding. Could it be that the newly discovered microRNA controls the activity of GAMYB?

Just in the last few weeks we've found that miR159 does

indeed target the mRNA encoding GAMYB. When both miR159 and GAMYB mRNA are present in the same cells, the mRNA is cleaved at the site of complementarity between the two. Furthermore, the sequence of miR159 is found in a variety of different plant species. It's found in thale-cress, in tobacco, and even in barley. These plants, although they once had a common ancestor, have been evolving in separate lineages for millions upon millions of years. If they share nucleotide sequences, those sequences must be doing something important to the life of plants in general. So miR159 could be expected to have a universally important role in the life of plants.

And now, to our great excitement, we've found that the amount of miR159 in plants is controlled by the growth hormone gibberellin. That gibberellin regulates the amounts of the miR159 by overcoming a repression of miR159 levels imposed by DELLAs. The possibility that DELLAs might repress the levels of a microRNA was something we hadn't previously considered, and the finding that they do is unexpected. A source of surprise. The thing that is particularly exciting about this observation is that it is a first. The first time that the levels of a plant microRNA have been shown to be promoted by a hormone.

Furthermore, the miR159 provides a level of 'homeostatic' regulation. It was already known that gibberellin promotes the production of GAMYB mRNA, the transcription of the *GAMYB* gene. Now we have found that gibberellin also inhibits GAMYB activity, by promoting the destruction of GAMYB mRNA via miR159. A subtle balance of promoting and inhibiting activities determines the final level of GAMYB activity.

This is all very well, but begs an important question. Can a developmental function be identified? Can it be shown that miR159 actually does something that affects the growth and development of plants? We've managed to answer these questions

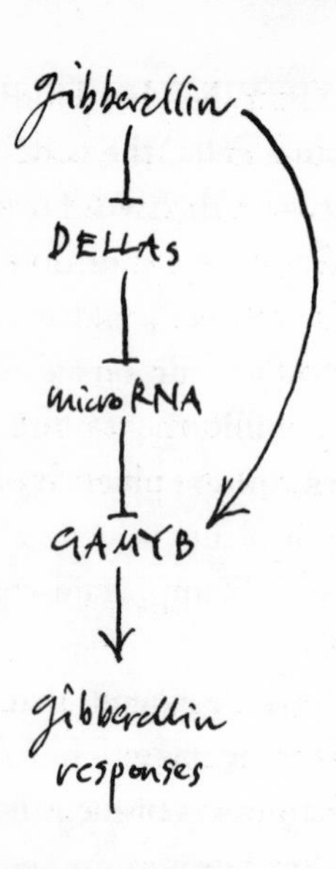

Gibberellin promotes GAMYB activity by directly stimulating the production of mRNA encoding GAMYB. In addition, gibberellin inhibits DELLAs, which themselves inhibit microRNA (miR159) levels. Thus gibberellin promotes miR159 levels, thus reducing the levels of mRNA encoding GAMYB. So, via different routes, gibberellin both stimulates and represses the levels of GAMYB-encoding mRNAs.

in the last few days by making plants that have elevated levels of miR159. This is done by driving its expression from a highly active promoter, thus making the levels of miR159 higher than they are in normal plants. As expected, the higher levels cause a reduction in the levels of mRNA that encodes GAMYB. But, more excitingly, we've found that the development of plants with increased levels of miR159 is indeed altered. These plants flower later than normal plants do.

In some ways this result isn't unexpected. We already knew that the flowering time of thale-cress is dependent on gibberellin, and there was previous evidence that GAMYB plays a key role in that process. Accordingly, reductions in the level of the mRNA encoding GAMYB would be expected to delay flowering. But the thing that is really new about this new observation is that we've shown that variation in levels of the miR159 can itself affect flowering time. Thus the miR159, and the effect it has on GAMYB activity, have developmental significance.

There is something mysterious about microRNAs. They are new on the scene. Not yet fully assimilated into our models of things. Reminiscent of 'dark matter' in cosmology. The micro-RNAs exist. They have the capacity to affect gene expression and hence to regulate development. But it's still unclear to what extent these tiny pieces of RNA actually do the things they might do. Nevertheless, it seems likely that they will turn out to be an essential determinant of the shapes and forms of the plants we find in the natural world.

Tuesday 19th October

A grey fog covers everything this morning. Orange beacons of turning leaves shine through it. A pervasive stillness.

It has just occurred to me that *excitement* and *exciting* are words I'm using with increasing frequency in these notes. Largely, I think, because I can't think of any other words to accurately represent the particular state of mind that accompanies the realisation of something new. The flashing of the light that is associated with the knowledge that some new experimental result or some sudden spark of idea marks progress in the development of insight. Yet the individual uniqueness of such moments is undoubtedly lost in the homogeneity of their description as being 'exciting'. And the experience of these moments of vision is of course highly pleasurable. The desire for that experience to be repeated a significant driver of the process of scientific enquiry.

What is it that gives these moments their peculiar intensity? There is of course the vindication of prediction, the completion of puzzle, the solution of a long-standing problem, a mystery sleuthed. These are all pleasurable in themselves. But do not, I think, provide a total explanation. No. The real excitement comes from the widening of vision. When suddenly, one more piece of the world comes into view. From now on, whenever I use those common words *excitement*, *exciting*, they will be invested with a sense of the individual uniqueness of the moment that they describe.

Wednesday 20th October

This morning, it was still dark when I got out of bed. Although it's noticeably more cold, there has been no frost. A cobweb, seen through my study window in the returning light, had its structure highlighted by the dew. And now, still later, the glorious golden light again harmonises with the colours of the autumn leaves. In addition to the oak, lime, and beech, the hazels are beginning to turn: yellowing leaves seen beyond the dew-greyed lawn.

At this moment it seems to me that there's no clear distinction between the things that are alive and the things that are not. Of each of the trees before me through the window, most of what I see is dead. A DNA molecule isn't alive. Living things are made of the non-living. So if we hold life to be in any sense sacred, then non-living things must necessarily be included in that sphere. But who knows how I'll see this tomorrow?

The leaves of different trees change colour at different rates. The beech in the next-door garden that I first noted as tinged with orange at the end of July still has many of its leaves (more brown now than orange). But the leaves of other trees, like the sycamore, started to change long after that beech did, and are already mostly shed.

Later in the day. It's been cold, with passing showers. Some so heavy that they were strongly audible: first as a tattoo of drum-beat reports, then as a steady roaring. Each individual impact carrying great force. When the rain eased off, I went to look at the thale-cress seedlings in St Mary's. And most of them have been flattened. Squashed on to the ground, bashed over. Hypocotyls horizontal. I think that they may have been uprooted.

But four of the older seedlings are still upright. There have been several warmish days since last I visited, and consequent increases in the numbers of leaves comprising each rosette. I counted them, one has five, one six, another nine leaves (eight large ones and the beginnings of the ninth), each set in a spiral. Having survived the battering of today's rain, these plants will now be taking up the resultant water from the soil. I think they'll continue to grow.

Thursday 21st October

After all the heavy rain of yesterday and the night before, today we have clear blue skies, yellow light, and fierce wind. It bites through clothes, catches the leaves between its teeth and tears them off the trees. The oak is being stripped, and a swirling carpet is forming on the lawn.

Saturday 23rd October

This morning the light is pale through mist, there's wood-smoke in the air, and everywhere I look the changing colours of the leaves strike the eye: lemon-yellow fig leaves, beech leaves umber with green flecks, oak leaves hanging limp and browning. The wind giving a sense of pace to the autumnal transition.

The salt-growth paper, as previously predicted, is turning out to be particularly hard to write. So far I've arranged the text into sections, based around three figures:

Fig. 1a. Seedling growth is inhibited by salt, but less so the growth of seedlings lacking DELLA proteins. Fig. 1b. Salt causes an increase in the accumulation of DELLAs.

Fig. 2. Salt delays the transition from vegetative to flowering phase, but much less so in mutants lacking DELLAs.

Fig. 3. Mutants lacking DELLAs are less able to survive long periods of exposure to high salt than are normal plants. Thus DELLAs help plants to survive periods of adversity.

I do love these findings. They're so much to do with landscape. With the way in which the world outside the plant shapes growth and development. And the structure of the paper is clear. But the story is complex, with many strands. Not a single line. And I'm having great trouble with the rhythm of it. The words seem constantly to jar against the flow, like cross-waves slamming contrary to the forward momentum of a ship at sea. The plan is to submit the thing to *Science*, and I'm going to have to resolve all this if there's to be any hope of it getting published there. I think part of the difficulty is that this is not a conventional genetics/molecular-genetics paper. It's at a level of abstraction that's higher than such papers tend to be, identifies links across wider distances than is usually the case. This is novel. But its very novelty makes it harder to write. After a while of wrestling with these complexities I find that my mind gets less able to see the picture, that all becomes fuzz.

Whilst writing just now I suddenly saw myself in my mind's eye standing next to my luggage near open doors on a Circle Line tube train at Edgware Road station, returning from Singapore a few weeks ago. Why did it come then, unbidden, with nothing obvious to associate it with the other things in my mind at that moment? The mind is an astonishing thing, logically following particular lines of linked thought and yet simultaneously bombarded with the distraction of sensation and those transiently intense flashes of memory.

Sunday 24th October

Last night was mild, windy, and wet. And I couldn't sleep. Mind too active. Thoughts tangling and spinning around each other. The season is definitely getting into me. An unsettling mixture of excitement at the pace and unease about where we're heading.

This morning went running in order to rein myself in. The lovely smells of wet autumn leaves and soil. The russets and browns, golds, oranges, reds, and yellows of the leaves. All contributed to the restoration of serenity.

Later I went to St Mary's. Those thale-cress seedlings that I'd seen flattened by the rain are now dead. They lie flat on the ground. Wilted, leaves shaped to the crumbs of soil to which they stick. However, the four that remained upright after the storms of last Wednesday are still growing.

Monday 25th October

Strong winds in the night, and several heavy showers. This morning many more leaves are down. The beech at the far end of the garden retains just a few, and those that remain on the sycamore are battered remnants: crumpled and torn, their shapes stark against the background of a shining grey sky. And apparently there's another storm gathering in the Atlantic, of greater severity than anything so far this autumn. It will be here sometime on Wednesday.

Tuesday 26th October

How to describe the keening of the blackbirds at dusk? Urgent? Acid-sharp stabbings of anxiety at the passing of the light? All this in my ears as I look down in the fading light at the remaining thale-cress seedlings. They are still there, those surviving four. Still growing.

There is something that I'm aware is coming increasingly to the fore in these notes. Yet something that I've also been skirting around. As if trying to avoid dealing with it face-on. And I'm particularly aware of it now, as I look down on those seedlings so recently sprung up from the surface of the grave. It is that there's a component of my feeling for them which has elements of the religious. Actually, this component has been evident since the beginning of this life-cycle study. But in recent months it's become ever stronger, colours my vision.

What is this religious feeling? I find it so hard to express, to pin down. But I think one aspect to it is the quality of *significance*. That somehow those seedlings, and the phenomena associated with them – growth and so on – have a significance that extends beyond what is actually observed. A significance that reaches out into the world and to the universe beyond. And this sense of significance is of course a self-affirming thing. It makes the seedlings seem more a part of the world than before, and the resultant sense of unity is itself something with religious resonance.

Yet there are problems here. It is of course me that is imbuing the seedlings with extended significance. It could be argued that significance is not a quality inherent to them but one that I impose. And as a scientist, I admit that there is a part of me that is queasy about such impositions. Furthermore, there's a jarring between the 'scientific' position and the 'religious' one. Because if we are within a unity that has greater meaning than the division (or sum) of its parts, it becomes difficult to know how representative of reality is an understanding gained from the isolation of objects. So what to do? I think that where I am with this today is that science works. It gives us a picture of the world, without which our vision would be immeasurably the poorer. But also that I do retain a sense of the religious.

Wednesday 27th October

The expected storm isn't very powerful here. Although it's been quite severe in Ireland. Floods in Bantry apparently. The seedlings are still fine.

Much of today spent writing the gravitropism paper. Making it into a vehicle of concise expression. Tightening it, making its parts connect. It's getting to be pretty good. This one's also for *Science*. Will they go for it? Who knows? But I think it has a chance. After all, identifying the thing that enables the different auxin levels to cause the different growth rates that cause root gravitropism *is* important.

Sunday 31st October

In one sense the first morning of winter. The clocks put forward in the night. But after the mist cleared, there was bright light between the clouds.

Ran to the cathedral and back and watched the light shape itself in blocks around the shower clouds. The colours of sky and leaves brilliant. Orange, scarlet, vermillion, shining with dark brightness. They glitter. Pierce and enter. Steer, subvert the mind. Gloriousness edged with apprehension.

A few days ago I saw an old man in the street. His head hung heavily from his neck, his grey skin slimed with sweat. His eyes tired, small-pupilled, pained. And I'm well down that path. Yet I keep going. We all know that we're heading that way, but mostly we ignore the unstoppable. In the bath after my run I looked at my legs. Not very elegant. Pimply. Not shapely either. But they work fine. Forty minutes to loop around the cathedral, to see its spire and weather-cock poking into the clouds, and then back home, drawing the city within my string.

I'm suddenly thinking that those remaining seedlings are vulnerable. It's so late in the year, and they each have only a few leaves in their rosettes. Now the increasing cold will slow them further. I fear that they might not have enough meat on them. That they are too delicate to survive the worst of the winter.

NOVEMBER

Wednesday 3rd November

THIS morning, on the way to work, my sight was struck by a sapling *Ginkgo biloba* on a grassy verge. I cycle past this tree most days on the way to work and have not previously noticed it. It was the colours that did it. A small, round beacon-ball of glowing leaves on the top of a bare, stick-thin trunk. Such striking brightness, as if phosphorescent – a brilliant lemony yellow.

Yesterday I had great fun. Spent almost the whole day working on the gravitropism paper. At a stage that I always enjoy: tightening it, making it concise, but also putting in flashes of light that point up important things. At times I felt that I was sculpting it: tapping off splinters, chipping away roughness to reveal the form and detail within the stone. Still a long way to go with it before submission. There will be more polishings and discussions on improvements. But more and more I have the feeling that it exists. That it's an entity with inherent solidity.

Friday 5th November

And now I'm swinging back again. Yesterday I reread and found that gravitopism paper lacking in so many ways. Full of strands that don't connect, observations poorly described. It's tempting to throw all the bits of it up in the air and see how they fall: perhaps I'll see a better pattern in the new heap. But no, it won't do. I'll revise yet again what I've got. It's funny how every time I write a paper I go through the same things. Get frustrated by the process, knowing all the time that the very forwardses and backwardses that generate that frustration are essential to the development of the final form.

But it's a fine, still morning. Blue sky. High wispy strands of white clouds.

The leaves, seen from my study window, are now astounding in their colours. The beech shining rust-brown in the sun. Those remaining few on the oak bronze or copper, the rest fallen. The form of the oak now mostly of visible twiglets, twigs, branches, and trunk, the bare wood being the clearest thing about it. The hazels still retain their leaves: yellow, some yellow-edged with green centres. The lime too: yellow, more yellow-brown than the hazels, some with centres blackened by the spread of fungus, a yellow and black dappled pattern.

Periodically a leaf falls, gliding from the lime to the ground. In a slow screwing or zig-zag motion. And on the recently raked lawn a spotted, random distribution of leaves. Not yet a thick enough covering to hide the grass again.

Later – to St Mary's in the dusk. The horse chestnuts are now naked, save for a few shrivelled remnants. Almost all the leaves have fallen, and the ground is covered. I had to lift one to see the thale-cress seedlings hidden below. But those seedlings *are* still growing. Perhaps young plants now rather than seedlings. A few more leaves added to the spiral of the rosette. The unseasonable mildness has helped. Here we are in early November and still there has been very little frost – no sustained periods of real cold.

Saturday 6th November

Just back from a fireworks party. Such airborne flashes, screechings, and bangs. The brilliant, instantaneous illumination of the surrounding skeletons of trees. The fleeting shadows, the sudden shapes of colour in the sky. I can write about it in ways we all recognise. As a record of passing experience, an attempt to capture the sense of evanescence, the second-by-second change and mutability of perception. These things are shared.

And although science too deals with a wonder in phenomena, it's difficult to write about it with such resonance. The images often seem dull, have a durability that lacks glitter. And the writing doesn't capture or restate shared experience. These are problems.

Sunday 7th November

THE DECLINE AND FALL OF LEAVES

This afternoon, Alice and I went to Wheatfen wood. For me to think more about the gravitropism paper. For her to gather some autumn leaves for a project at school. We shuffled under the trees in our wellingtons. Slanted beams of light cut through the branches.

Towards its end, autumn seems to have more more velocity than the other seasons. To have more of a trajectory, plunging into winter. And today there was excitement in the momentum of it. In the flashes of light, the racing sky of blue patches and grey cloud driven west by an easterly wind. In the sunbeams that briefly warmed me through my coat.

We saw a bracket fungus tiered on a fallen trunk. The texture of the bracket and the lump of the bark so similar that it was as if the one was a swollen or distorted form of the other. As Alice said, it was hard to know where bark ended and fungus began. A metamorphosis.

Then deeper into the wood in search of different leaves. Amongst the intensifying smell of leaf decay, acid, sweet, pungent, pricking at the nostrils. And looking at the leaves still on the trees I was aware again of the speed of change: the green transforming to gold, orange, amber, yellow, red, and brown. I was moved by the beauty of it all. Sometimes, when a tree is caught in a gust of wind and shimmering leaves begin suddenly to fall through the air, it can take your breath away.

Before they fall, these dying leaves provide the material for new life. Next year's leaves will be made from them. The macromolecules – the proteins, lipids, and carbohydrates that comprise the cells of the old leaves – are broken down into their smaller constituent molecules and absorbed into the trunk. Retaining the resources that would otherwise be lost to the tree when the leaf is shed. Resources that will be used to build new leaves in the coming spring. The whole process analogous to the browning of the leaves of the thale-cress plant in St Mary's a few months ago.

We walked further into the wood. Stopped before a young beech. Peered up at the limbs in their graceful reaching curve towards the sun, the leaves graded and patched in colours from green through yellow to gold and then brown. In places amongst the canopy there were gaps where leaves had once been, now marked out by bare twigs only. I reached up, took a leaf from one of the lower branches, handed it to Alice. The veins were clear, a strong mid-vein constituting a central rib that ran from base to tip, alternating ribs that ran from the mid-vein out at an angle to the edge of the blade. Yet smaller veins that came off the alternating ribs at regular intervals, and these branched again to fill the space that was left.

Alice noticed that the tissue around the alternating veins was green, but that between these green fingers, in those parts of the leaf that were furthest from the main veins, were patches of brown. I explained to her about the nutrients leaving the leaf along the veins on their way to the trunk. She said that it was as though the tree was sucking up green goodies, using the veins as straws. And as she said this I was reminded that still I don't really know where life begins and ends.

Tuesday 9th November

I've spent the whole day on the gravitropism paper. Rewritten it from beginning to end. And now I've just reread and what can I say? Simply that it won't do. The writing lacks clarity. It's *still* clumsy, lacks elegance.

Wednesday 10th November

ON ASTONISHMENT

There is much in life that seems ordinary. We get up, take the childen to school, go to work, come home again, go to bed. Doing and seeing the same things. All predictable. But just occasionally, there's the splintered flash of the sublime. At such a moment, for instance, one might remember that the world is the minutest speck in an infinite void. In those brief moments it's clear that the world and everything in it is extraordinary.

And there's a related paradox in science. The establishment of scientific truth depends on the reproducibility of observation. Once established, reality is predictable. There is of course the excitement of the initial discovery. But quickly that discovery, if proven, becomes familiar. An accepted truth, unremarkable. There's a tendency for knowledge to become uninteresting once understood.

I'm increasingly aware of the need to oppose these tendencies. I will see those thale-cress seedlings in St Mary's as the extraordinary things they are. And I will continue to be astonished both by the known world, and by what our experiments reveal. When these things are clearly seen as being astonishing, their significance, that gravity of significance I wrote about a week or so ago, becomes more apparent. More appropriate, I think.

With the realisation of the relevance of astonishment I feel a sense of relief. As though a weight I'd previously been unaware of carrying has vanished. From now on I'll look at the world in joy and surprise.

Friday 12th November

ON AUXIN CARRIERS

At last I think it's forming. As always the writing goes in peaks and troughs. I'm on a high now. The abstract is at last how I wanted it: succinct, sharp. And I've hugely improved one whole section that just wasn't properly realised before, a section about auxin transporters. One of the sub-plots of the paper is the analysis of auxin movement in DELLA-deficient roots. This movement is due to the action of specific carrier proteins. For example, the root auxin-fountain is driven by carriers that are at the bottom of cells (as defined by the direction of flow). They are called 'efflux' carriers: they remove auxin from one cell so it can enter the next. Mutants that lack efflux carriers display reduced gravitropism, because the auxin flow no longer works.

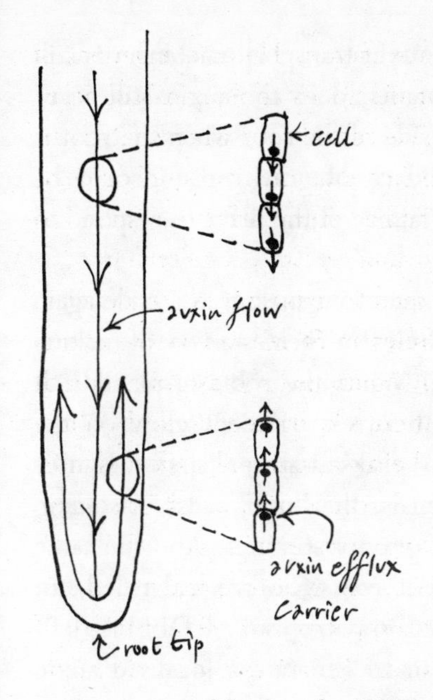

Efflux carriers are on the bottom of cells in central vasculature, where auxin travels down the root, and in the top of peripheral cells, where auxin flows back up the root.

It was possible that the DELLA-deficient mutants had impaired gravitropism because the flow of auxin was impaired, perhaps because DELLA deficiency somehow affected the efflux carriers. This possibility needed to be tested. So we did. We stained the cells of DELLA-deficient mutant roots with an indicator that highlights efflux carriers. And we found them. In exactly the same places as they are in normal plants. These images are so revealing, and my previous descriptive efforts just didn't capture them. But I've got it now, and the point is made: the impaired gravitropism of DELLA-deficient mutants isn't due to lack of auxin-transporting proteins.

Saturday 13th November

ON AUXIN ACCUMULATION IN GRAVITROPISM

The wind has moved round to the north-west. Last night it coursed through the trees in roaring gusts. Today it's cold, and tonight there may be a proper frost, the first of the season.

So the next question is: if the auxin-transport machinery is still present in DELLA-deficient mutants, does the auxin still accumulate in the cells of the lower side of the root when the root is turned? If it doesn't, then the lack of gravitropism could be explained by this rather than by failure of the plant to respond to the auxin accumulation.

We were able to answer this question, making use once again of the green fluorescent protein (GFP). There's a way of making GFP shine in response to auxin. When the roots of plants that have this capacity are turned, there's a transient glow on the lower side of the root. This glow peaks at around six hours after turning, then decays, and signals the brief pulse of auxin accumulation in response to the gravity stimulus. And the same thing happens in DELLA-deficient roots. So we had ruled out another possibility. The impaired gravitropism of DELLA-deficient roots isn't because they fail to generate a localised auxin accumulation in lower-side cells.

Sunday 14th November

I have *too* much to do – grant proposal on the go, papers to write, a research group to inspire and manage, committees to sit on, the list is endless. However, I have a feeling of purpose in my work that was lacking at the beginning of the year. I'm excited. Although overwhelmed, I have a sense of direction that enables me to steer a path through these things.

A quick trip out to Wheatfen. The autumnal ruin develops. The reeds yellowing and stiffening. Their colours echo the trees. The fen a sombre expanse. Mistiness and low cloud above it. Chill in the air. A lonely coot-call. All is still.

Monday 15th November

ON THE SPLITTING OF THINGS

There's a moan of wind in the trees. It reminds me of the pines that surround the house in Ireland. And that in turn reminds me of a recital one July evening a few years ago in the nearby church of St James's, Durrus. The Bach D minor chaconne played in the summer's heat by Catherine Leonard on an unaccompanied violin. Her sweating fingers slipping and adding to the poignant fragility of the music's realisation. Amidst a stink of sewage, a thick fetor rising from the mud-flats of the nearby estuary at low tide. The music made potent in that atmosphere: its stretched quality, its sense of strained to the limit. A cathedral constructed of the taughtest wire. All thrown into relief by the stench of mortality.

It occurred to me that the separation, the boxing off of things, isn't a characteristic of science alone. That our entire culture is built on it: we constantly classify, divide, separate one thing from another. Perhaps to block off the view, ignore the distasteful, the incomprehensible, the frightening, whatever. We partition ourselves from the unpalatable realities of our biology: that we shit, that we kill to eat, that we will die.

Tuesday 16th November

It's mild. Light a dim grey. November in its dismal decrepitude: decaying leaves dancing about. The acceptance that things are coming to their end. And this sense of impending closure has caused me to reread some of these notes, going back to May and June. My overall impression is that there's something useful here. The speed of the writing reflective of energy. The text sketchy, the oil as if still wet, unvarnished. Of course there are things that protrude. And the capturing of half-finished thoughts makes some of it incompletely realised. There are ideas that would have been discarded if the writing were more 'finished'. Yet there's life in it. It does depict the progress of this year.

Wednesday 17th November

THE CONCLUSION OF THE GRAVITROPISM STORY

So to the denouement. The final twist in the gravitropism tale. It's simple really. DELLA-deficient mutants display reduced root-gravitropism. Yet they accumulate auxin on the lower side of gravitationally stimulated seedlings just like normal plants do. So the thing that must have gone wrong in the mutant plants is the way in which they respond to the gravity-induced auxin accumulation.

And the experimental results are consistent with this conclusion. If you artificially recreate the gravitationally induced auxin accumulation by growing roots on a medium containing a high concentration of added auxin, the growth of normal roots is inhibited. This is not the same as you get from gravity, of course. The extra auxin is now on both sides of the root. Rather than curving, the growth of the root as a whole slows down. Both sides do what one side does when it curves in response to gravity. When the same experiment is done with DELLA-deficient roots, they carry on growing. They fail to respond to the auxin.

And so the case is made. Roots need DELLAs if they are to respond appropriately to gravity-induced auxin accumulations. It's a significant advance in understanding. But I still don't think I've captured the essence of it in the paper. Whenever I reread it I feel insecure, that it's not fully realised.

Saturday 20th November

It's cold. Last night there was a severe frost, the first of any real impact this winter. We revved up the heating in the house – but I still didn't sleep very well. I'm just not good at winter. I get cold in the night and wake up with a feeling of pressure on my chest, bumping heart, sweat-drenched. Last night it wasn't that bad. But neither do I like the weight of the blankets, the dryness of the heat. I can't get it right. It's one or the other.

But I'm in good spirits. Yesterday I finally got the gravitropism paper to form itself. It's a question of pace and level. For a paper to work it has to do so on several levels, from that of the overall to that of the detailed particular. A paper that lacks definition on these several levels seems flat, featureless. And the pace has to work simultaneously at each of those levels. Often it isn't obvious how to do it. One just has to keep on trying new things, and then suddenly you know that you're there. I reached that point yesterday.

I'll reread it again tomorrow, doubtless will tweak and prink at the odd word or phrase, tighten down a few rattling screws. But it won't change much. And then, on Monday or Tuesday, it's the covering letter to the editors at *Science* (which must be compelling!), and submission.

To return to my recent musings on the idea of significance. This idea has become more central to my thought during the course of the year. To begin with, following the thale-cress plant and subsequent seedlings in St Mary's, made me connect the growth of plants with the seasons more closely. Then there was our discovery that DELLAs enable the graduated restraint of growth in response to changes in the environment. So growth is as much a property of the environment as it is of plants. A sense of oneness, of the world having inherent unity, is a natural progression from this point. And that this unity is of overwhelming importance. Sacred. The best word I can think of to sum up my feelings about it.

Sacred. A word with religious resonance, of course. But I don't use it here to imply God-given, I can't go that far. Rather, I want to propose that we should view the world with awe and humility. Our actions increasingly threaten the fundamental properties and stability of the earth. Would we change our ways if the idea of the world as sacred were more current?

Sunday 21st November

I went this morning to St Mary's to find that the grave has been tended. This time the work has been done well. The gravel weeded and neatly raked. There's no sign of the seedlings. All are gone.

At first it seemed so very disappointing. I'd looked forward to seeing if those seedlings would make it through the winter. Then I remembered that there will probably be some dormant seeds still lying in the soil, waiting for spring.

And now I'm back at home, reflecting, and can see that the story is already told. The purpose with which I began these notes, to the extent that I then understood it, is at least partially fullfilled. And my excitement with our science is rekindled. So this is the end, or at least part of the end, of these notes.

Tuesday 23rd November

It is mild again, blown in yesterday on a steady west wind, shreds of clouds scudding fast across the sky, a sense of energy and excitement in the air. And yesterday, *at last*, we submitted the gravitropism paper to *Science*. It's all so amazingly electronic in these days of the computer. The files containing text, figures, supplementary material, etc., all moved via the internet from my computer to the *Science* editorial office. At the click of a mouse it pings its way across the Atlantic. So, now it's done and now we wait.

There are several stages to this. First stage: will the editors think the paper is worth considering? If so, on to the second stage: the paper will be sent out for in-depth review. If the reviewers like it, they're still sure to ask for changes. Rewriting a passage, perhaps, even further experiments. So the rewrite is the third stage and then it's back to the editors for the final judgement. Acceptance or rejection. And if it's rejected, at any stage, then I'll just have to rewrite it and send it somewhere else.

DECEMBER

Friday 3rd December

HAVEN'T written in these notes for days. Over the last week I've been completely focused on the writing of a grant proposal (deadline today and finished at last). And now I'm enjoying the relief of having it completed and temporarily out of my hands. Yesterday the weather was alchemical: blue, gold, and orange from sunlight refracting/reflecting through/from the clouds. Some of the view of it went into the final sharpening phases of writing the proposal, I think.

This morning got up to the kind of piercing frost and fog that penetrates clothes, cycled through flat expanses of white to work.

And my anticipation is mounting incrementally. Every day I check the e-mail, and still there's nothing from *Science*. It's more than ten days now since I submitted the gravitropism manuscript. Surely they will not, after this time, return it (refusing it) unreviewed. I think they must have sent it out for peer-review. The first hurdle crossed.

Sunday, 5th December

There is a scientific strand in these notes that needs to be drawn together. The importance of transcription factors. I have shown how, to give some instances, the response to cold, the timing of flowering, the identity of trichomes and petals, growth; all are controlled by transcription factors. By proteins which, whilst encoded by genes, return to the nucleus to influence the activity of other genes.

Let's take a step back here. Look at this from a broader perspective. It's generally understood that genes affect the growth of organisms. Many people know of the work of Gregor Mendel, how he showed, for example, that a mutation in a single gene could make a plant grow short. The prevailing common view that has developed from such classical experiments is that genes act independently, that gene x, say, discretely performs role y. The fact that all of the developmental processes listed above are controlled by transcription factors points out the over-simplicity of this view. Instead, the biology of organisms is driven by communication between genes, by genes influencing the activities of other genes.

I have described a few cases of what we know of simple linear relationships between genes, where gene A influences the activity of gene B. But our current knowledge is superficial. It is clear that the relationships between the different genes in the genome are vastly more complex than this. That they interact in patterns of massive and subtle complexity, in ways as fascinating but intangible as the interior workings of the brain. The development of organisms is thus not the product of individual genes, but of the interactions between the genes of an organism's genome. By genes working in harmony. Influencing and being influenced by each other's actions. And the science required to understand it all will have to be complementary. Equally subtle. Equally complex.

Friday 10th December

I went back to St Mary's this morning. But there's no sign of thale-cress seedlings. And it's unlikely that any remaining seeds will germinate now. It's too cold for that. If there are seeds still dormant in the soil, they'll wait until next spring.

Still no reply from *Science*. What will the reviewers make of our paper? I've been wondering what weaknesses they might identify. But every published scientific paper has holes in it, leaves questions unanswered. That's part of the process.

Yesterday I resumed work on the salt paper. It needs a lot of work – tightening, focusing, etc. It's very like the gravitropism paper was at this stage – flat, all on one level. It needs the resonances emphasised and the depths plumbed. Presumably I'll get there eventually. I'm hugely excited about the conclusions of this paper. They extend in so many different directions. Already there are indications that the slowing of growth that occurs when plants become diseased is dependent on DELLA-restraint. I'm becoming increasingly convinced that with our simple salt experiment, our naïve attempt to recreate the harshness of the salt-marsh, we've revealed something that is fundamental to the ways in which plants allow the outside world to control the rates at which they grow.

Friday, 17th December

I've just been reading *That's the Way I See It* by David Hockney. It contains a reproduction of one of his photo-collages, an image of a tree from the Jardins du Luxembourg in Paris that simultaneously incorporates the other trees lining the intersecting paths of the garden into the picture. For me this collage encapsulates some of the preoccupations of these notes. It is constructed from individual snapshots of different moments of time and degrees of scale. Close-ups of bark, dying leaves, pigeons on the path. Larger-scale views of entire trees shrunk with distance. The result is a synthesis of 'tree' that is at once near, and tapering to the horizon. A representation of what we really see with our momentary focusings of vision, our pastings of transient images into some coherent whole. And the collage integrates the different scales, encompassing single leaves, entire trees, the lawns and paths of the garden within the one image.

A conceptual integration of scales, that somehow allows us to see life simultaneously at both the visible and invisible levels, would, I think, enrich our experience of the world.

Saturday 18th December

ON SUBJECT AND OBJECT

It's cold. Wind in the east. Sky brilliant blue, brittle and crystalline. Snow is forecast. Still the gravitropism paper is with *Science*. And still I struggle with the salt paper.

I've been thinking more about the ability of science to represent reality. It has certainly given us insights that we wouldn't have without it. Visions, vistas, representations. And for that reason, if for nothing else, it's a shame that these ideas aren't more easily appreciated by non-specialists. That their relevance to our daily lives can seem obscure. The writing of these notes has been, at least in part, an attempt to make a few of the images that science has generated more visible to others.

But this isn't my particular preoccupation today. I'm actually more concerned with the nature of the insights themselves. More and more I'm coming to the view that they are distorted. Created with the artifice of objectivity. It's a conceit. We say that there is an object, and that we are separated from it, observing it. But the picture is necessarily incomplete. It is inherently stretched and strained, with some features exaggerated, and others omitted. As with a novel or painting, it is not an absolute representation of reality.

And I'm beginning to think that this extends beyond science. That we're culturally conditioned to view the world in terms of subject–object. To draw a dividing line between what is inside, 'self', and what is outside, 'non-self'. Our whole lives conducted against this constant background. The subject–object split a premise, a given, that is engrained within our thought. Automatic, unappreciated.

So our picture of the world is necessarily inaccurate. How can it be otherwise when we are actually both subject and object?

How, then, to improve our pictures? Make them a more faithful representation of reality? I don't know. This is too big a problem for me to grapple with. But I can think of some things that might help.

First, an acceptance that the images from science are distorted, some more than others. And at the same time an appreciation that science has value because it helps us to see, however inaccurately, things that otherwise could not have been seen at all. Second, a general adoption of the view that the world is a whole. That it is part of us and we are part of it. See ourselves as part of something sacred. Exercise restraint.

This is the last time I'll write these notes. The story of the plant is ended. I'm more settled in my mind. This notebook has served its purpose. I have direction. Have shifted our science away from a focus on the hidden secrets of plant growth to a broader vision that considers simultaneously the plant in the world and the world in the plant.

AFTERWORD

15th July 2005

Today I went back to St Mary's. Although I've often returned to Wheatfen in the months since December 2004, it was the first time since then that I'd returned to the churchyard. It remains much as I remember it. A place of peace amongst the graves and brambles. A haven of shelter in the lee of towering horse chestnuts.

The grave itself is completely barren. Kempt. Nothing grows amongst the gravel. And despite an intensive search, I was unable to find a thale-cress plant anywhere else in the churchyard. It seems that the plants I studied here were migrants. Brought in from outside. Perhaps as seeds clinging to the hem of someone's coat. As if what I recorded were the final stages in the life of a colony.

And what of the manuscripts I was completing and submitting for publication towards the end of 2004? The gravitropism paper was reviewed by *Science* and then rejected. But the rejection came with an invitation to resubmit once we had obtained further data. In the expectation that the results that we obtain will provide a more substantial test of the relationship between root gravitropism and DELLAs. These further experiments are currently in progress, and we should resubmit quite soon. The salt-growth paper, as predicted, and despite its novelty, was very difficult to write. Went through many phases of drafts and revision. Finally we submitted it, again to *Science*, and it's currently being reviewed.

I remain keenly excited by what we found in the experiments

we did in the latter half of 2004. And I'm beginning to see the growth of plants as a metaphor for our times. The DELLAs being the pivot of responsiveness, the agents that restrain the rate of growth to a degree consonant with the conditions within which plants find themselves. Lacking DELLAs, a plant becomes insensitive, brash, a fast-liver that is unable to exercise appropriate restraint, and that dies young. The appropriateness of restraint is a message that we ourselves need to heed.

Whilst writing these concluding words, I'm looking out of the study window to the oak beyond. The soft coo of wood-pigeons is audible from the chimney-stack above my head. And I'm recalling that the contents of all our cells – oak, wood pigeons, myself – are derived by descent from some starting gel of cytoplasm that formed itself so very long ago in the waters of the sea. Despite the vastness of the time that has passed since then, we all carry within us some remnant of that first stirring of life.

Glossary

Words in bold in each entry indicate cross-references within the glossary.

ABC model: explains how the distribution of three different **transcription factor** activities (A,B,C) over the four different **whorls** from which a flower is constructed results in the formation of **sepals, petals, stamens** and **carpels**

***AGAMOUS*/AGAMOUS: gene/transcription factor** specifying floral organ identity (see **ABC model**)

amino acid: the fundamental unit from which **proteins** are built, comprising twenty different types (eg. **arginine**)

amyloplast: starch grain-containing **organelle** found in root cap cells that sediments in response to gravity and thus regulates root gravitropism

anthocyanin: purple pigment found in cells of flowers, leaves, stems etc.

***APETALLAI*/APETALLAI: gene/transcription factor** specifying floral organ identity (see **ABC model**), also **inflorescence meristem** identity

arginine: an **amino acid**

asparagine: an **amino acid**

auxin: a plant growth hormone that travels from top to bottom (from cell to cell) of the plant via a specific transport mechanism

auxin efflux carrier: protein localised in the bottom membrance of cells that facilitates removal of auxin from those cells (and hence transfer of auxin to the cell beneath)

bases: the fundamental units of **DNA** and **RNA**, in **DNA** comprising four different types: A (adenine), C (cytosine), G (guanine) and T (thymine)

***BOOSTER (B)*/BOOSTER (B): gene/transcription factor** regulating **anthocyanin** pigmentation in maize cells

carbohydrate: molecule of general formula $C_x(H_2O)_y$; examples are starch, sugars, **cellulose**

carbon dioxide: molecule of formula CO_2; a gas in the atmosphere. Plants use carbon dioxide to construct **carbohydrates** and other complex organic compounds during **photosynthesis**

carpel: the innermost organ of the flower. In thale-cress the **gynoecium** forms from the fusion of two carpels and contains the ovaries. The top (**pollen**-receptive) surface of the gynoecium is the **stigma**, below this the **style**. Following fertilisation of ovaries by the **sperm nucleus** of the pollen, the ovaries develop into seeds and the gynoecium forms the fruit (pod)

***CAULIFLOWER*/CAULIFLOWER: gene/transcription factor** regulating **inflorescence meristem** identity

***CBF*/CBF: gene/transcription factor** regulating plant response to cold

cell: fundamental unit of living organisms. In plants comprises a **nucleus**, **cytoplasm**, surrounding **membrane**, and **cell wall**. Plant cells usually also contain a central fluid-filled **vacuole**

cell division: the process whereby one **cell** divides to make two

cell expansion: the process of cell enlargement

cell wall: rigid structure surrounding the **membrane** of plant cells. Made from cellulose and pectin

cellulose: fundamental constituent of **cell wall** comprising long chain **macromolecule** made of glucose (sugar) molecules. Molecules are laid down in bundles to make fibres

chlorophyll: green pigment responsible for light capture during **photosynthesis**

chromosome: the structure into which the **DNA** is packaged within the **nucleus**. The thale-cress **genome (DNA)** is distributed between five chromosomes

codon: set of three **bases** in **DNA** (and **RNA**) signifying a particular **amino acid** component in the **polypeptide** chain from which **proteins** are constructed

***CONSTANS*/CONSTANS: gene/transcription factor** regulating flowering time in response to photoperiod (length of the daylight phase of the twenty-four dark/light cycle)

cotyledons: embryonic (seed) leaves. Site of storage of seed reserves. The first two leaves visible on germinating seedings, distinct from the 'true' leaves generated by the **shoot meristem** following germination

cytoplasm: the contents of the cell other than the **nucleus**. A suspension of **macromolecules** in water, also larger-scale structural features (**organelles**) defined by surrounding **membranes**

***D8*:** a maize dwarf **mutant**. The gene mutated in this **mutant** encodes the maize **DELLA** protein

DELLA/DELLAs: a family of related proteins that restrain the growth of plants. The thale-cress **genome** contains **genes** encoding five distinct DELLAs: **GAI**, **RGA**, **RGL1**, **RGL2** and **RGL3**

diploid: thale-cress plants are said to be 'diploid' because their cells contain two copies of the **genome**; one from the female parent, the other from the male

dominant: the (**diploid**) cells of thale-cress contain two copies of each **gene**. If one or other of these copies is a **mutant** gene, then it might be **dominant** or **recessive** to the normal gene. If dominant, the plant will exhibit the characteristics conferred by the mutant gene. If **recessive**, the plant will exhibit the characteristics conferred by the normal gene

DNA: the material from which **genes** are constructed; a long double-stranded **macromolecule** consisting of a sequence of **bases**

embryo: the beginnings of the plant as contained within the seed; consists of **cotyledons**, **hypocotyl**, embryonic root, **shoot meristem** and **root meristem**

enzyme: a type of **protein** that catalyses (facilitates; speeds rate of) specific chemical reactions within the **metabolism** of the **cell**

epidermis: outermost layer of **cells**, some of which become specialised to form **guard cells** or **trichomes**

ethylene: molecule of formula C_2H_2; acts as a hormone that regulates plant growth

excision (of transposon): the process via which a **transposon** jumps out of a segment of **DNA** into which it is inserted

***FLOWERING LOCUS C*(*FLC*)/FLC:** gene/protein that controls flowering

FRIGIDA: gene that controls flowering in response to cold

***GAI*/GAI:** *GAI* is the 'original', 'normal' form of the gene that encodes the GAI **DELLA** protein

gai*/gai:** the *gai* gene is a mutant gene derived from ***GAI, and encodes a mutant (gai) form of the **GAI** protein. *gai* confers dwarfism, and is **dominant** over the ***GAI*** gene because plants carrying both *GAI* and *gai* are dwarf rather than tall

gai-d*:** another mutant form of ***GAI, obtained following radiation treatment of ***gai*** (see text). The *gai-d* form encodes no protein, and is thus **recessive** to both ***GAI*** and ***gai*** forms of the gene

gai-t6*:** another mutant form of ***GAI, this time containing a **transposon** inserted into the ***gai*** **open reading frame** (see text). The *gai-t6* form encodes no protein, and is thus **recessive** to both ***GAI*** and ***gai*** forms of the gene

gamete: germ cell, the reproductive cells of male (**sperm nucleus**) and female (egg) that fuse to form the **zygote**

***GAMYB*/GAMYB: gene/transcription factor** that controls some responses to **gibberellin**

gene: unit of genetic information, made of **DNA**, individual gene written in italics, e.g. *CBF*; often consisting of a region where gene expression (eg. rate of **transcription**) is controlled (the **promoter** region) and a region that encodes protein (the **open reading frame**)

genome: the complete set of **genes** in the **DNA** of the **nucleus** of **cells**

germination: the process during which a mature seed **embryo** resumes growth, bursts out of the seed-coat, and becomes a viable seedling

GFP-DELLA: a 'fusion' protein consisting of the **green fluorescent protein** with a **DELLA** protein fused to it

gibberellin: plant growth hormone

***GLABRA1*/GLABRA1: gene/transcription factor** that regulates **trichome** development

glume: flower-case in grasses eg. maize

gravitropism: the process via which plant organs orientate growth with respect to the gravitational vector. Thale-cress seedling roots are positively gravitropic (grow towards the centre of gravity), shoots are negatively gravitropic (grow away from the centre of gravity)

green fluorescent protein (GFP): a protein that flouresces green when exposed to ultra-violet light, and can thus be used as a marker, allowing microscopic visualisation (imaging) of its location within living cells

guard cells: crescent-shaped **cells** of the **epidermis** that occur in pairs on either side of pores, **turgor**-driven changes in shape of these cells regulate the flow of gas and water vapour through the pores and thus into or out of the plant body

gynoecium: female reproductive organ of the flower, in thale-cress formed by the fusion of two **carpels**

hypocotyl: the 'stem' of the **embryo**. Elongates during **germination** to lift **shoot meristem** and **cotyledons** above the surface of the soil

insertion (of transposon): the process via which a **transposon** jumps into (integrates itself within) a **DNA** segment

***LEAFY*/LEAFY: gene/transcription factor** that regulates the transition from **vegetative meristem** to **inflorescence meristem**

leucine: an **amino acid**

macromolecule: very large molecules (e.g. **DNA**, **protein**, **carbohydrate)** usually polymer chains built from unit molecules (e.g. **bases**, **amino acids**, **sugars**)

membrane: extremely thin layer of fat and **protein** that surrounds all **cells** and **organelles**

'Mendelian' ratio: the classic 3:1 (or 1:2:1) ratio first discovered by Gregor Mendel. When a plant carrying one **dominant** and one **recessive** form of a gene is self-pollinated then three-quarters of the progeny in the next generation exhibit the characteristics conferred by the **dominant** form

meristem: a focus of prototypical **cells** that maintain themselves via **cell**

division and at the same time acts as a source of the cells from which the rest of the plant body is built

floral meristem: meristem from which floral organs are derived

inflorescence meristem: meristem from which **floral meristems** and inflorescence stem are derived

root meristem: meristem from which root cells are derived

vegetative (shoot) meristem: meristem from which leaves and stem are derived

metabolism: the chemical processes that occur within cells, involving the destruction (catabolism, reduction of molecular complexity) and construction (anabolism, increase of molecular complexity) of molecules, the energy released by catabolism and harvested during **photosynthesis** being used to drive anabolism

microRNA: very short (21 **base**) **RNA** molecules that target specific **mRNA** molecules for destruction

mRNA: see **RNA**

mutation: an alteration in **DNA** sequence that is transmitted from one generation to the next

mutant (gene): a gene that carries a **mutation** and is thus either destroyed or changed in the way in which it works. For example, ***gai*** is a mutant gene that encodes an altered protein, whilst ***gai-t6*** is a destroyed gene that does not encode a protein

mutant (plant): a plant that carries a **mutant gene** and that may have visibly altered characteristics as a result. For example, a *gai* mutant plant is dwarfed because it carries the mutant *gai* gene

nucleus: the **organelle** within the **cell** that contains the **genes**

open reaing frame (ORF): region of **DNA** sequence (part of a **gene**) that encodes **protein**

organelle: sub-region of cell bounded by a **membrane** and usually specialised for a specific task, e.g. **nucleus**, chloroplast (site of light-harvesting by **chlorophyll**)

parenchyma: characteristic cells/tissue of stem/root, often permeated by inter-cellular spaces containing air

petal: characteristic organ of the flower, usually found in the second **whorl**

photosynthesis: the synthesis of organic **macromolecules** from water and **carbon dioxide** using energy absorbed by **chlorophyll** from sunlight

phototropin: a blue-light photoreceptor that is especially involved in phototropism (movement of plant organs towards the light)

phytochrome: red/far-red light reversible photoreceptor

phytochrome-GFP: fusion protein consisting of **phytochrome** fused to **green fluorescent protein**

pollen: source of **sperm nucleus** (male **gamete**) during floral pollination, also contains a vegetative nucleus (responsible for growth of pollen tube down the **stigma/style**)

polymerase chain reaction (PCR): a method for rapid amplification of specific **DNA** segments

polypeptide: a segment of **amino acid** sequence. **Proteins** are large polypeptides

promoter: the region of the **gene** that controls rate of **mRNA transcription**

proteasome: a microscopic sub-cellular chamber containing **protein**-digesting **enzymes**. Proteins marked by poly-**ubiquitin** enter this chamber and are then destroyed by the enzymes

protein: encoded by a **gene**, a polymer of **amino acids**, the 'active' entity that the gene encodes: **enzymes**, **transcription factors** are proteins. Individual proteins are written in normal script, e.g. the CBF protein is encoded by the *CBF* gene (via a **mRNA** intermediate)

recessive: see **dominant**

'relief of restraint': the model that explains how **DELLAs** control the growth of plants. Essentially, DELLAs repress growth, whilst gibberellin promotes growth by causing the destruction of DELLAs

***RGA*/RGA:** *RGA* is the 'original', 'normal' form of the gene that encodes the RGA **DELLA** protein

rga-24: a mutant form of ***RGA*** that encodes no protein, and is thus **recessive** to **RGA**

***RGL1*/RGL1:** *RGL1* is the 'normal' form of the gene that encodes the RGL1 **DELLA** protein

***RGL2*/RGL2:** *RGL2* is the 'normal' form of the gene that encodes the RGL2 **DELLA** protein

***RGL3*/RGL3:** *RGL3* is the 'normal' form of the gene that encodes the RGL3 **DELLA** protein

***Rht*:** a wheat dwarf **mutant** (*Rht* stands for *R*educed *h*eigh*t*). The gene mutated in this **mutant** encodes the wheat **DELLA** protein. The *Rht* **mutation** contributes to the high yields of modern 'green revolution' wheat varieties

rosette: see **vegetative rosette**

RNA: single-stranded molecule carrying a sequence of **bases**, usually made by **transcription** of a segment of **DNA**

mRNA: (messenger RNA) RNA transcript of a segment of a **gene** containing an **open reading frame**

SCF complex: a multi-**protein enzyme** that adds polymers of **ubiquitin** to proteins, thus targetting them for destruction in the **proteasome**

sepal: floral organs usually found in the outermost **whorl** of the flower

serine: an **amino acid**

sperm nucleus: the male **gamete**
spongy mesophyll: characteristic cell/tissue layer of leaves, often permeated by inter-cellular spaces containing air
stamens: male organs of a flower, carry **pollen**-forming anthers at their tips
stigma/style: stigma is the top surface, the style the upper portions of the fused **carpels** of the flower. Germination of **pollen** on the stigma results in formation of a pollen tube which penetrates and passes along the style on the way to the ovaries
stop codon: a codon that signifies no amino acid and thus terminates an **open reading frame** and the **protein** (or **polypeptide**) that it encodes
transcription: the formation of **RNA** (usually **mRNA**) that corresponds in sequence to a segment of **DNA** (usually an **open reading frame**)
transcription factor: a **protein** that controls the rate of **transcription** of a **gene** by interacting with the promoter
translation: the formation of a **protein** (or **polypeptide**) that reflects the sequence of the **open reading frame** of a **gene** as previously **transcribed** into **mRNA**
transposase: the **enzyme** that promotes the **insertion** and **excision** of a **transposon**
transposon: 'mobile' DNA, a short segment of **DNA** that can remove itself (**excision**) and reintegrate (**insertion**) within a larger segment of DNA
transposon-tagging: a method that uses a **transposon** for the isolation of specific genes (see text)
trichome: hair cell found in the **epidermis**
turgor: pressure generated when the rigid **cell wall** opposes the increase in volume that accompanies absorption of water by **vacuole** and **cytoplasm**
ubiquitin: a small **protein** that, when added to other proteins in chains of 'polyubiquitin', marks those proteins for destruction in the **proteasome**
vacuole: fluid-filled space (an **organelle** bound by a **membrane**) within the **cytoplasm** of plant **cells**. Can be relatively large, comprise most of the volume of the cell
vegetative rosette: the relatively flattened spiral of leaves that forms around the central stem of the thale-cress plant prior to flowering
vessels: tube-like structures constructed from cells placed end-to-end
xylem: vessels that conduct water and salts throughout the plant
phloem: vessels that conduct sugars and **proteins** throughout the plant
whorl: circular region that defines a set of identical floral organs (e.g. **petals**)
***WUSCHEL*/WUSCHEL: gene/transcription factor** that regulates the size of the **vegetative meristem**
zygote: the product of fertilisation of the male and female gametes. The first cell, from which the embryo, and eventually the mature plant body, is entirely derived

Acknowledgements

I'd like to express my gratitude for the help of many people in the writing of this book. Firstly, to my scientific colleagues. I am indebted to Michael Freeling. My time in his laboratory (at the University of California, Berkeley) was crucial to the development of my ability to think scientifically. My own research group made many of the discoveries described in the book. Past and present members are: Patrick Achard, Tahar Ait-ali, Liz Alvey, Mary Anderson, Marie Bradley, Pierre Carol, Rachel Carol (née Cowling), Andy Chapple, John Cowl, Thierry Desnos, Hiroshi Ezura, Barbara Fleck, Xiangdong Fu, Llewelyn Hynes, Kathryn King, Jinrong Peng, Pilar Puente, Carley Rands, Donald Richards and Yuki Yasumura. Their individiual contributions are reflected in our scientific publications, and I thank them all. Some of the discoveries described in the book resulted from collaboration with scientists at the John Innes Centre, but outside of my own immediate group: amongst these were Katrien Devos, John Flintham, Mike Gale, George Murphy, John Snape and Tony Worland. Additionally, I gratefully acknowledge collaborations with members of the laboratories of David Baulcombe, Jonathan Jones, Klaus Palme, Jinrong Peng, Thomas Moritz and Dominique Van Der Straeten.

I am grateful to the John Innes Centre for giving me the latitude to write this book, and in particular to my friends and colleagues Enrico Coen and Keith Roberts for much valued advice and encouragement.

I am a first-time author, and getting this book to publication has been a fascinating journey. Many thanks to Alison and Stephen Cobb, Liz Fidlon and Anthony Harris for essential suggestions during the initial phases. To my superb and energetic agent Felicity Bryan I give thanks and admiration. Especial thanks to all at Bloomsbury: above all to Bill Swainson for his gentle, patient and steady editorial hand. Also, many thanks to Alexandra Pringle for her enthusiasm for the project and her comments on early versions of the text. The final manuscript was skilfully sharpened by the insightful and precise copy-editing of Andrea Belloli. The feeling of the book is immeasurably enhanced by Polly Napper's lovely and precise illustrations and diagrams, I thank both her and Will Webb, who created the design for the jacket using Polly's drawings.

As is evident, much of the thinking about this book was done at Wheatfen broad, the nature reserve managed by the Ted Ellis Trust. The warden and trust do a superb job in their stewardship of this important natural habitat. It is a haven of peace and inspiration.

I would like to thank my mother and father, Muriel and David Harberd, for the roles they played in shaping my abilities to create anything at all, let alone write a book. Above all, I thank my own family, both for putting up with the physical and mental absences that writing this book entailed, and for their help. Alice and Jack provided inspiration. And most importantly, my wife, Jess, who has encouraged me throughout, acted as sounding-board, as first reader and critic, and who believed that I could write something worth reading at times when I didn't.

All of the above contributed to the merits of the book. Any faults are my own.

N.H.

Norwich, November 2005

A Note On The Author

Nicholas Harberd is one of the world's leading plant biologists. He currently directs a research team at the John Innes Centre in Norwich and is Honorary Professor in the Department of Biological Sciences at the University of East Anglia. In 2008 he will become Sibthorpian Professor of Plant Sciences and Professorial Fellow at St John's College, University of Oxford. He is the author of numerous scientific papers, has published in the leading international scientific journals *Nature* and *Science*, and is the father of Alice and Jack.

A Note On The Type

The text of this book is set in Perpetua. This typeface is an adaptation of a style of letter that had been popularized for monumental work in stone by Eric Gill. Large scale drawings by Gill were given to Charles Malin, a Parisian punch-cutter, and his hand cut punches were the basis for the font issued by Monotype. First used in a private translation called *The Passion of Perpetua and Felicity*, the italic was originally called Felicity.